Workbook for

HARMONIC PRACTICE

IN TONAL MUSIC

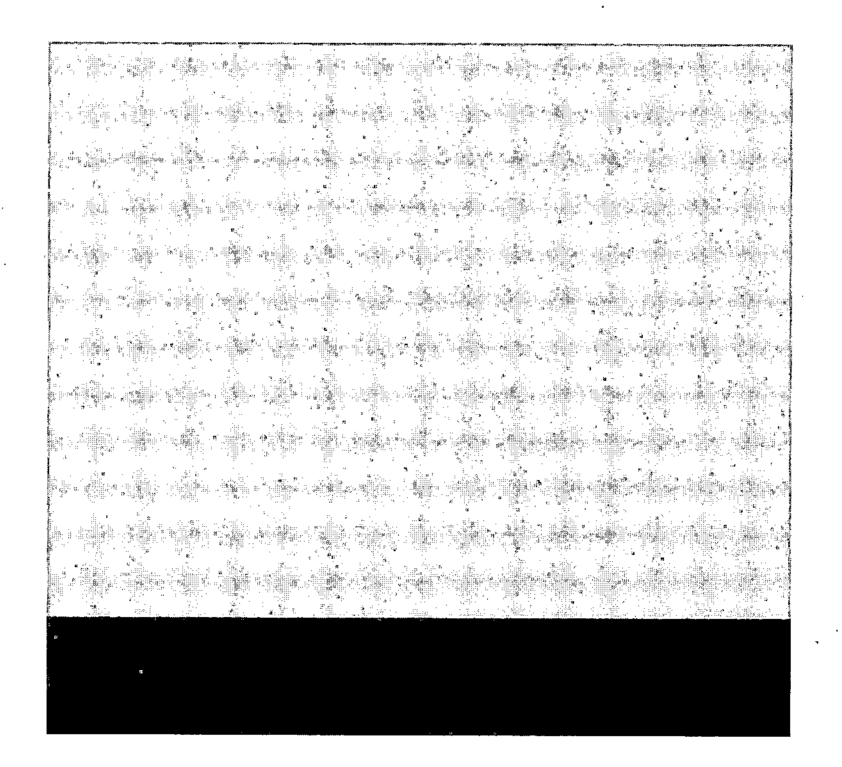

Workbook for

HARMONIC PRACTICE

IN TONAL MUSIC

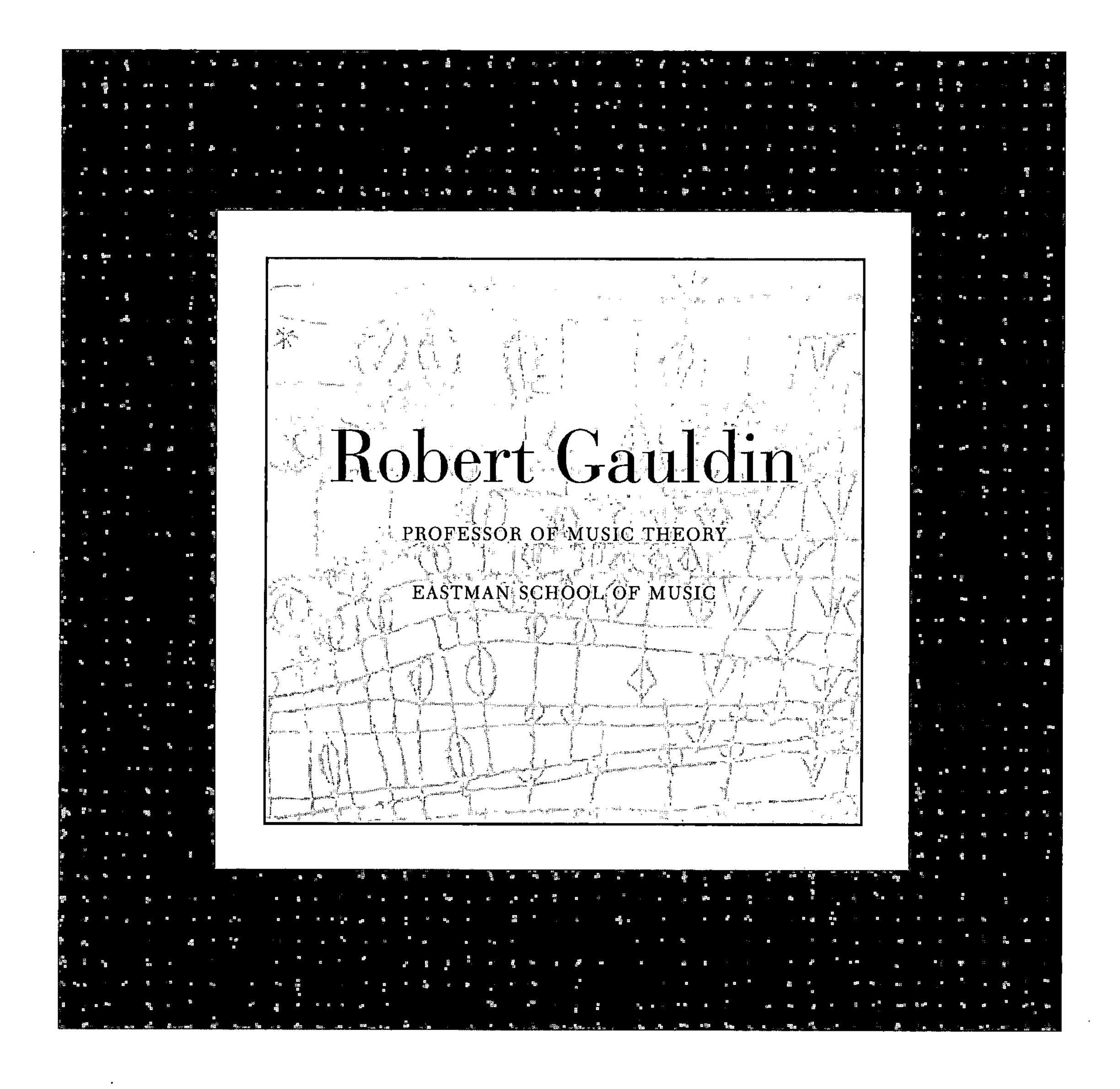

W · W · NORTON & COMPANY · NEW YORK · LONDON

ACKNOWLEDGMENTS

Desmond, "Take Five" (p. 128): Written by Paul Desmond. Used by arrangement with Desmond Music Co.

Kern, "The Way You Look Tonight" (p. 202): Words by Dorothy Fields. Music by Jerome Kern. Copyright © 1936 PolyGram International Publishing, Inc. and Aldi Music. Copyright Renewed. All Rights for Aldi Music Administered by The Songwriters Guild of America. International Coypright Secured. All Rights Reserved.

Rachmaninoff, Prelude in C♯ Minor, Op. 3, No. 2 (p. 189): Used by permission of Boosey & Hawkes, Inc.

Strauss, *Don Juan* (p. 162): By permission of C. F. Peters Corporation, New York.

Stravinsky, *Le Sacre du Printemps* (p. 53): © Copyright 1912, 1921 by Hawkes & Son (London) Ltd.; Copyright Renewed. Used by permission of Boosey & Hawkes, Inc.

Printed in the United States of America
First Edition

The text of this book is composed in New Caledonia, with the display set in Bauer Bodoni.
Composition by A-R Editions.
Manufacturing by Courier.
Book design by Jack Meserole.
Cover illustration: Paul Klee, "Ripe Harvest" © 1996 Artists Rights Society (ARS), New York/VG Bild-Kunst, Bonn.
Cover design by Kevin O'Neill.

ISBN 0-393-97075-2 (pbk.)

W. W. Norton & Company, Inc., 500 Fifth Avenue, New York, N.Y. 10110
http://www.wwnorton.com

W. W. Norton & Company Ltd., 10 Coptic Street, London WC1A 1PU

2 3 4 5 6 7 8 9 0

CONTENTS

PART III: CHROMATIC HARMONY

PART IV: ADVANCED CHROMATIC TECHNIQUES

KEYBOARD EXERCISES

PREFACE

This workbook provides the student with various types of written assignments and a series of keyboard exercises, which are included in the final section. Both types of assignments should be reinforced by an appropriate aural skills program that embraces sight singing as well as melodic and harmonic dictation.

THE WORKBOOK EXERCISES

The workbook assignments include short harmonic models to complete, figured-bass exercises (some not supplied with figured-bass numbers), melodies for harmonization, sequence-completion exercises, original composition projects, and excerpts for analysis. Comments and questions accompanying the analytical assignments can be taken up in class discussion.

For those chapters that focus on specific chord functions, we recommend that the instructor precede the assignments with oral and written drills that focus on spelling the new chords in various keys.

Although the number of assignments is sufficient to cover the main points of each chapter, the instructor should feel free to make up additional exercises or delete those that seem redundant, as the need arises. It is imperative that students be given enough assignments to master the particular topic discussed in each chapter.

With the exception of the five extended and sometimes demanding excerpts listed below, the excerpts for analysis from music literature are not recorded on the CD set. Instead, students are encouraged to acquaint themselves with the remaining excerpts at the keyboard, which provides the best possible method of absorbing the sound and structure of theoretical elements and concepts.

1. Schubert: March Militaire, Op. 51, No. 1 (Chapter 2)
2. Mozart: Symphony No. 36 in C major ("Linz"), III, Minuet (Chapter 23)
3. Bach: Prelude in C major from *Well-Tempered Clavier*, Book I (Chapter 23)
4. Schumann: "***" from *Album for the Young*, Op. 68, No. 21 (Chapter 30)
5. Wagner: "Liebestod" from *Tristan und Isolde*, Act III (Chapter 36)

The Keyboard Exercises

The keyboard exercises can serve as an intermediary step between the written assignments and ear-training drills, providing a more practical, performance-oriented approach to theory. These drills reflect the written assignments found in the workbook and therefore should be used in conjunction with them.

The exercises will familiarize the student with the sound of the various intervals, chords, and harmonic progressions in tonal music. Although primary emphasis is placed on playing the chords and harmonic progressions at the piano, the student should also keep in mind the melodic characteristics of the soprano and bass voices, heard both separately and in combination.

The keyboard exercises are arranged by chapter; there are no exercises for those chapters that do not require keyboard reinforcement.

Most of these exercises are not hard to play from a technical standpoint—especially those in the earlier chapters; usually, the problem is *what* to play (a function of the mind) rather than *how* (a function of the fingers).

PART ONE

CHAPTER 1

Pitch and Intervals

1 Write the indicated succession of melodic major and minor 2nds. Do not substitute an augmented prime (such as C–C♯) for a minor 2nd (C–D♭). Your concluding pitch should be D^4, as given.

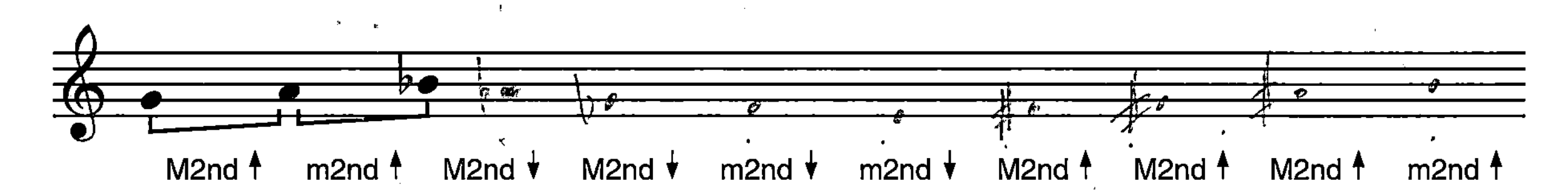

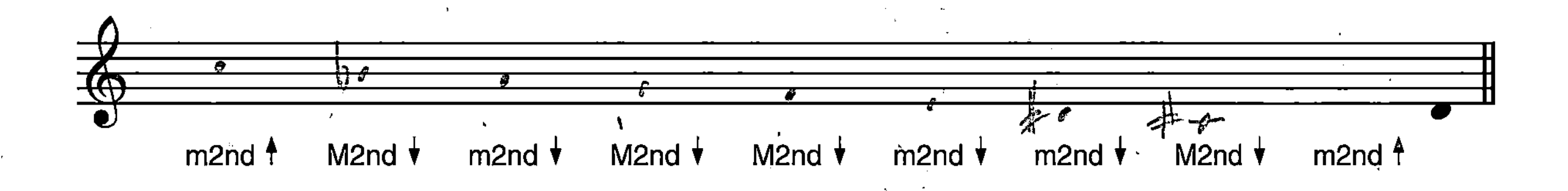

2 Study the succession of intervals in the opening measure. Then continue this established pattern in strict fashion to the final note, which should be $C\sharp^6$, as given.

3 Using *only* minor 3rds and major 3rds (no augmented 2nds or diminished 4ths), build an ascending succession of harmonic 3rds above C^2 in the bass clef so that it contains all twelve pitch classes with no enharmonic duplications. You may wish to work out your solution on another sheet and then copy the results here. Can you figure out some interval pattern that would avoid an extended "trial and error" approach?

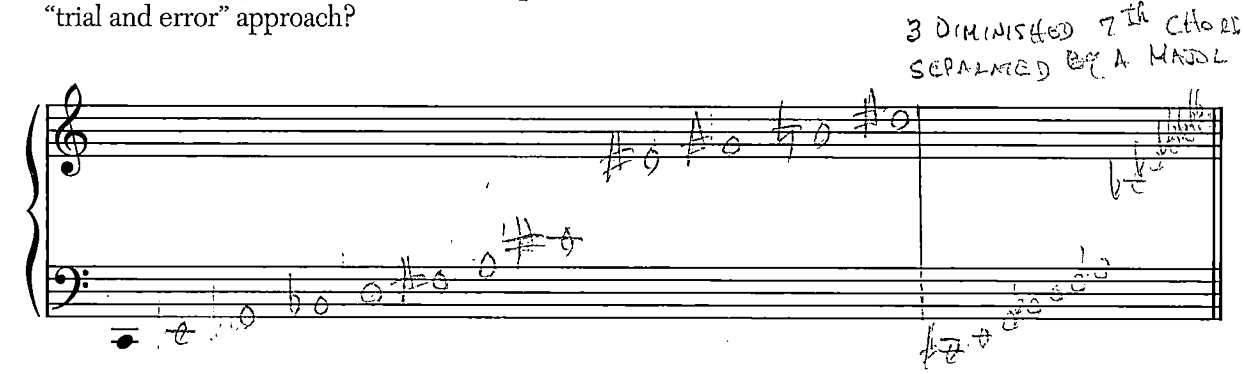

4 The following series of harmonic 6ths occurs in Chopin's "Double-Sixth" Etude, Op. 25, No. 8. Denote the type of 6th (major or minor) below the staff. Do you see recurring patterns? If so, bracket them.

5 **A.** Circle all examples of harmonic minor 7ths.

B. Circle only harmonic major 7ths.

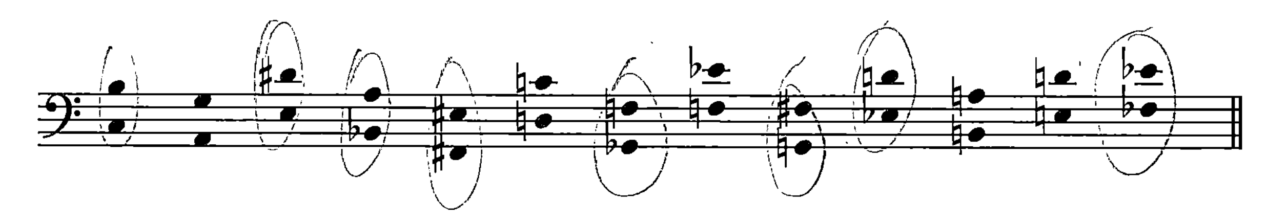

NAME ______________________________

6 The following set of pitches appears in Alban Berg's *Lyric Suite.*

A. Analyze each successive melodic interval in the spaces provided. The first two are done for you.

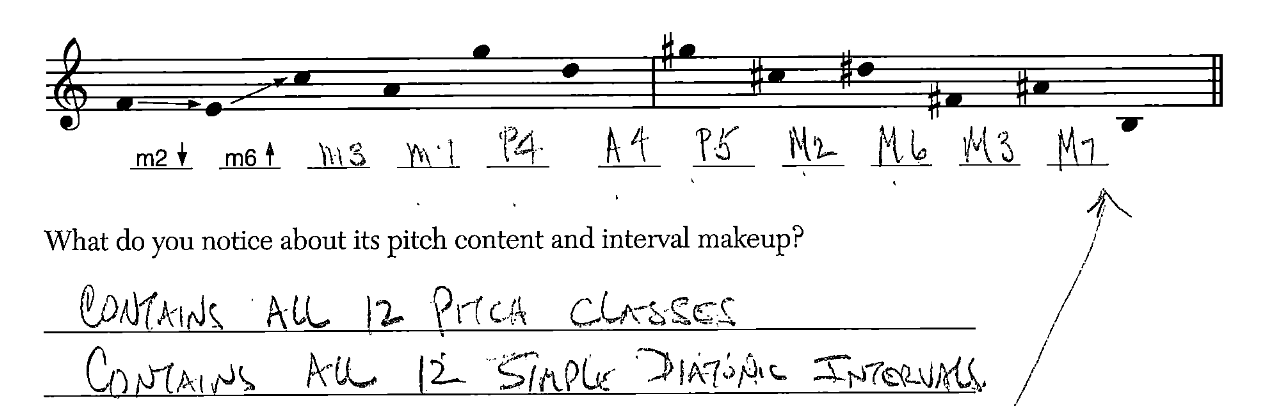

What do you notice about its pitch content and interval makeup?

CONTAINS ALL 12 PITCH CLASSES

CONTAINS ALL 12 SIMPLE DIATONIC INTERVALS

B. Now reverse the direction of each interval by writing its exact inverted or mirrored form (see the first three notes).

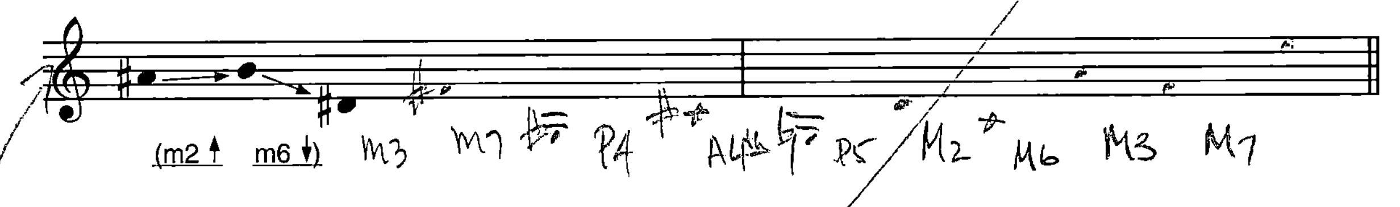

Compare each group of six pitch-classes with those of the original set; what has happened?

THE PITCH CLASSES IN THE FIRST SIX-NOTE GROUP (OR HEXACHORD) HAS EXCHANGED WITH THE PITCH CLASSES IN THE SECOND SIX-NOTE GROUP.

EACH SET IS AN INTERVALLIC INVERSION OF THE OTHER IN RETROGRADE MOTION

B: A♯ → B → D♯ = A: B → A♯ → F♯

m2 M6 M7 M3

7 Write the indicated diminished or augmented intervals above or below the given tones. Do not use enharmonic spellings.

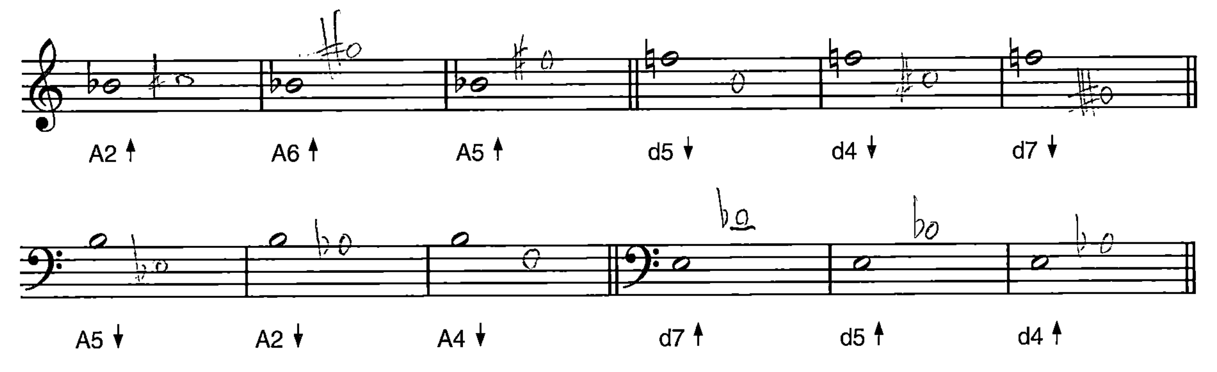

8 Identify the given harmonic interval and specify the number of half steps it contains. Then write its inverted form (perfect 5th = perfect 4th, etc.) and fill in the number of half steps it contains. The first example is done for you.

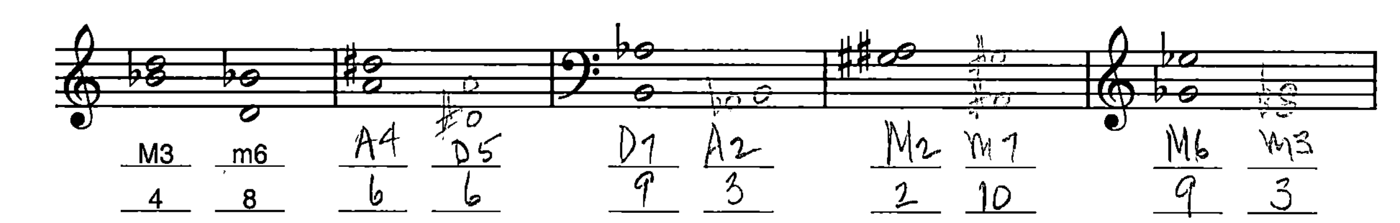

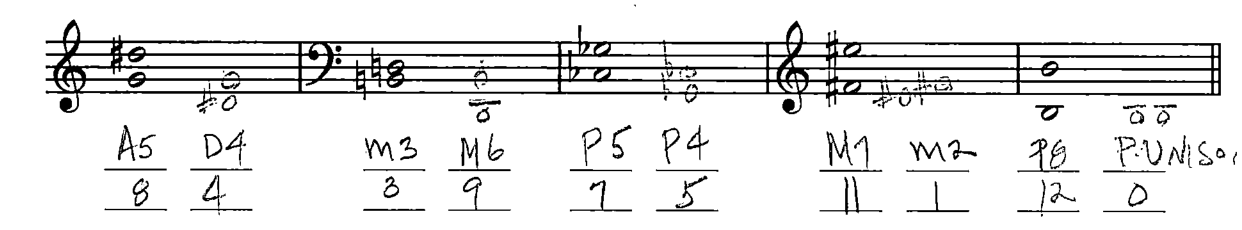

9 Circle all dissonant harmonic intervals in the two passages below and identify each one. Bracket any instances of two successive dissonances you find.

A. Each lower note is held for three upper notes, as indicated by the slurs.

B. Some notes—both upper and lower—are tied together.

10 As a class, try making a catalogue of tunes that begin with different melodic intervals, both ascending and descending. For instance, *The Star Spangled Banner* opens with a descending minor 3rd. For which intervals did you have trouble finding tunes?

NAME ________________________________

C H A P T E R 2

Rhythm and Meter I

BEAT, METER, AND RHYTHMIC NOTATION

1 Play or listen to the following passage on CD3. Now list those elements that contribute to the impression of (duple) meter.

1. DYNAMIC ACCENT ON FIRST BEATS, 2. SAME RHYTHMIC PATTERN IN FIRST THREE MEASURES, 3. BASS NOTE ON 1st BEATS, 4. EACH HARMONY LASTS APPROXIMATELY ONE MEASURE.

SCHUBERT: MARCHE MILITAIRE, OP. 51, NO. 1

2 Choose one of the metrical situations listed below and compose an original melody (about eight measures long) for a solo woodwind instrument. The choice of pitches for your melody is up to you, but your piece should convey the meter (or lack of it) that is listed. Perform your piece in class to see if your fellow students agree.

a. Compound duple meter.
b. Simple triple meter.
c. Compound quadruple meter.
d. Regular beat, but no sense of one prevailing meter.
e. No regular beat.

3 Analyze the following meter signatures, using the format given in the first example. In order to do this correctly you *must* take the tempo markings into consideration!

Moderate $\frac{2}{4}$	= $\frac{2}{♩}$		Allegretto ¢	=	
Adagio $\frac{9}{4}$	=		Grave $\frac{6}{8}$	=	
Vivace $\frac{12}{16}$	=		Allegro $\frac{9}{8}$	=	
Presto $\frac{3}{4}$	=				

4 **A.** Beam the following succession of eighth notes to correspond to metrical groupings in $\frac{3}{4}$ and $\frac{6}{8}$, converting all tied notes into the appropriate larger note values.

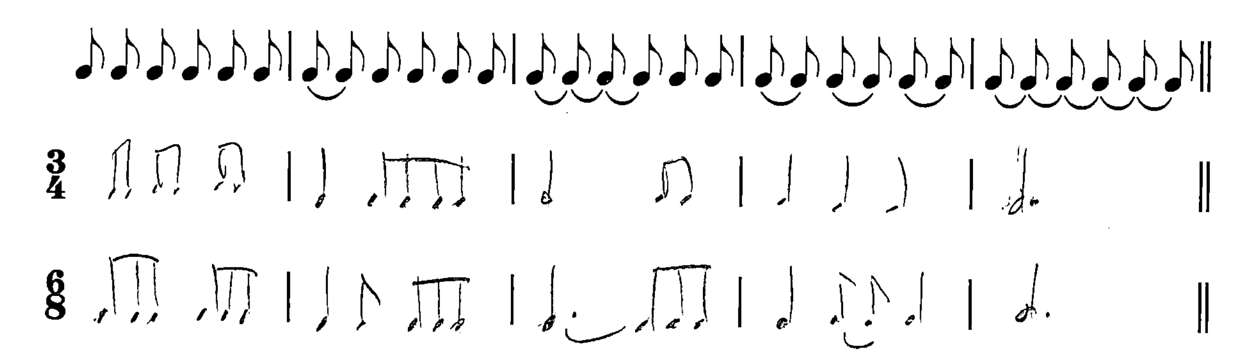

B. Regroup the following series of note values into measures of either $\frac{3}{2}$ or $\frac{4}{2}$, keeping the notes in the same order that is given. You may have to break up some of the note values if they tie over into the next bar.

Does this rhythmic succession suggest a stronger sense of triple or quadruple meter?

TRIPLE
____

Where do the barlines in the two examples align themselves?

LAST MEASURE

Can you give a reason why they do so?

FOUR OF 3/2 AND THREE MEASURES OF 4/2 CONTAIN THE SAME NUMBER OF BEATS.

5 Supply a meter signature that is appropriate for the rhythmic grouping in each measure below. In some cases, more than one signature is possible; see the first example.

6 Circle the notational errors in the following three rhythmic passages. There are a total of eight errors.

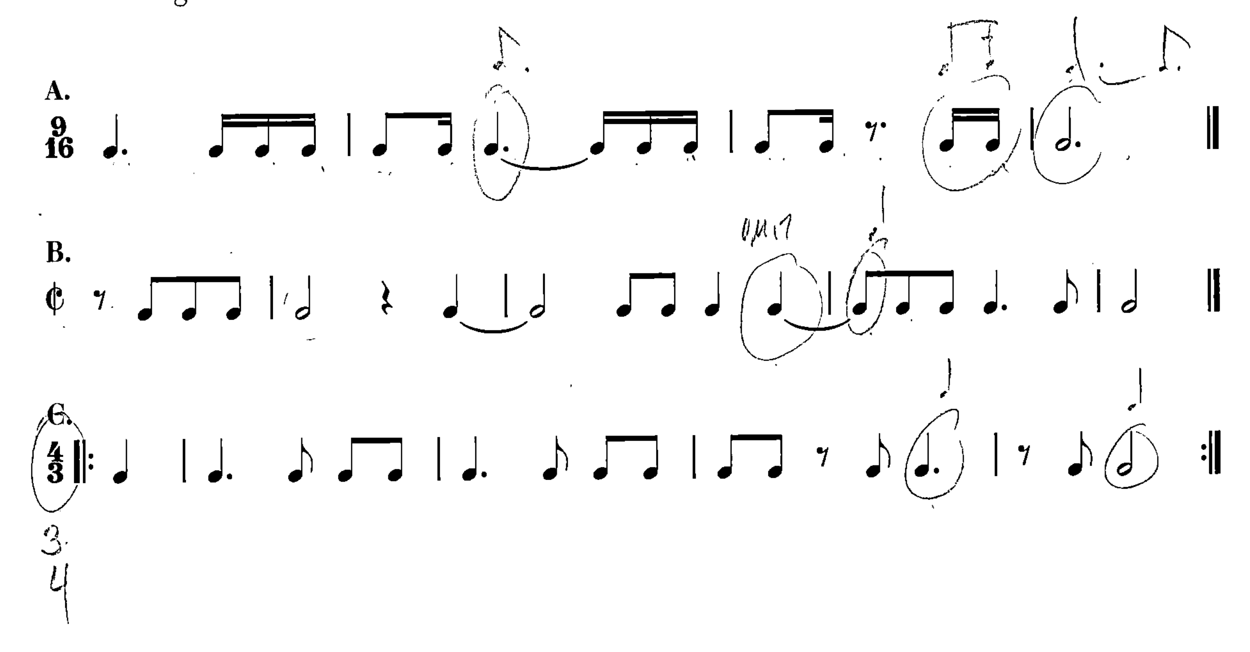

7 (Optional) As an exercise in transferring music accurately, carefully copy the first three pairs of staves from Example 3 on page 274 of the text to the staves that are provided here. Strive to reproduce every single mark as faithfully as possible. Completing this exercise will help you become proficient in notating and copying music, a task you will be called on to do in many exercises in this workbook.

NAME ______________________________

CHAPTER 3

Tonic, Scale, and Melody

1 Circle the tonic tone in each of the melodies below. List several factors that you feel establish that particular note as tonic. Be prepared to discuss these in class.

1. FREQUENT OCCURRENCE OF TONIC, ON 1st BEAT
2. FREQUENT USE OF 5th RELATIONSHIP
3. LEADING TONE RESOLVING TO TONIC

____

____

A. JOHANN STRAUSS: "TALES FROM THE VIENNA WOODS"

B. "Dreydl" (Jewish folk song)

C. "Amazing Grace" (hymn)

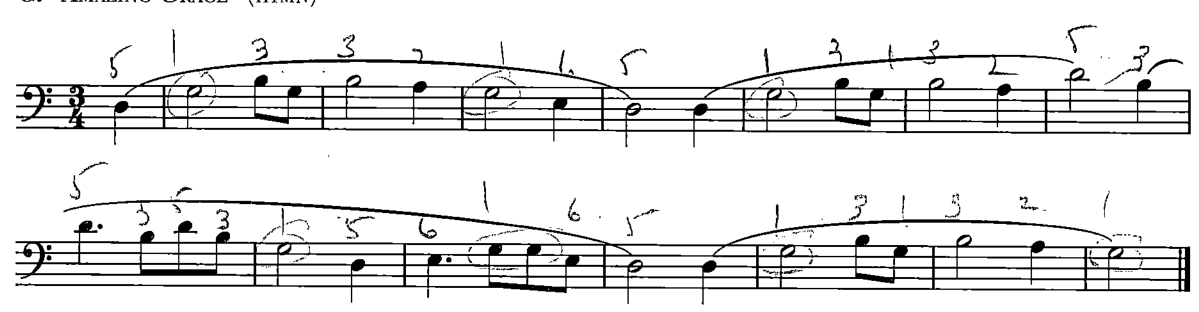

2 **A.** Given one specific pitch as a scale degree, write out the indicated scale, using accidentals but no key signature. Mark the successive intervals in each scale, using ‿ for a major 2nd, ^ for a minor 2nd, and ⌐¬ for an augmented 2nd.

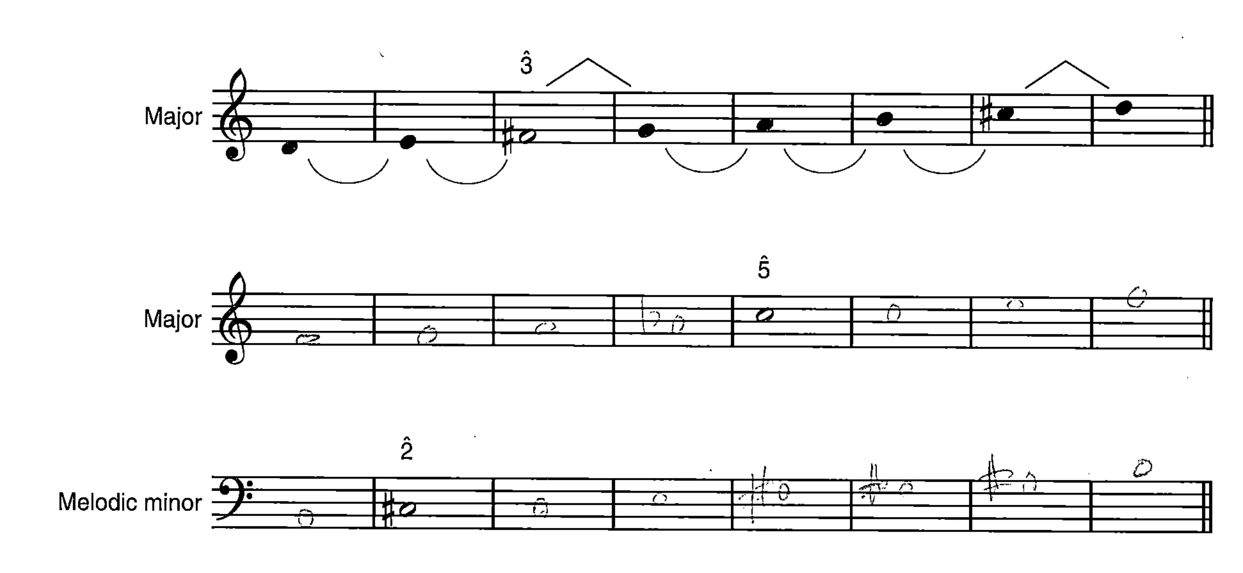

NAME ___________________________

B. Now go back and write in the scale-degree numbers above the notes in the tunes of Exercise 1. What do you observe about the last tune?

Pentatonic scale

____

3 In what scale(s) does each set of four notes occur?

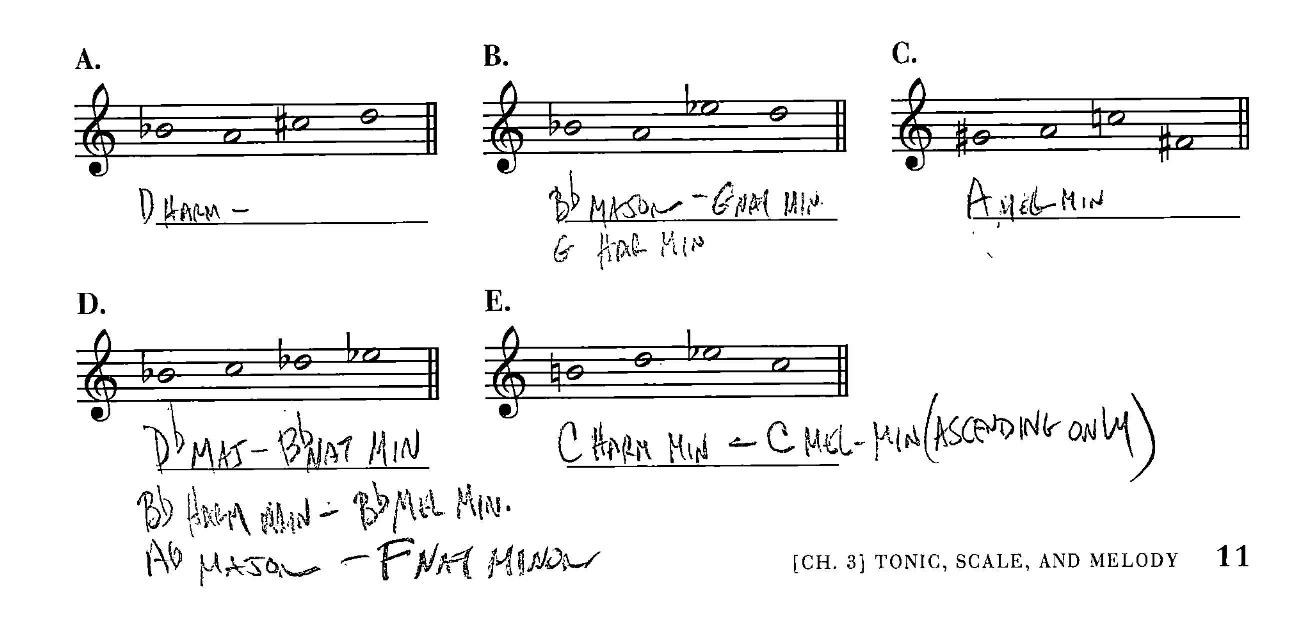

4 Write the key signatures for the following keys and their parallel major or minor keys.

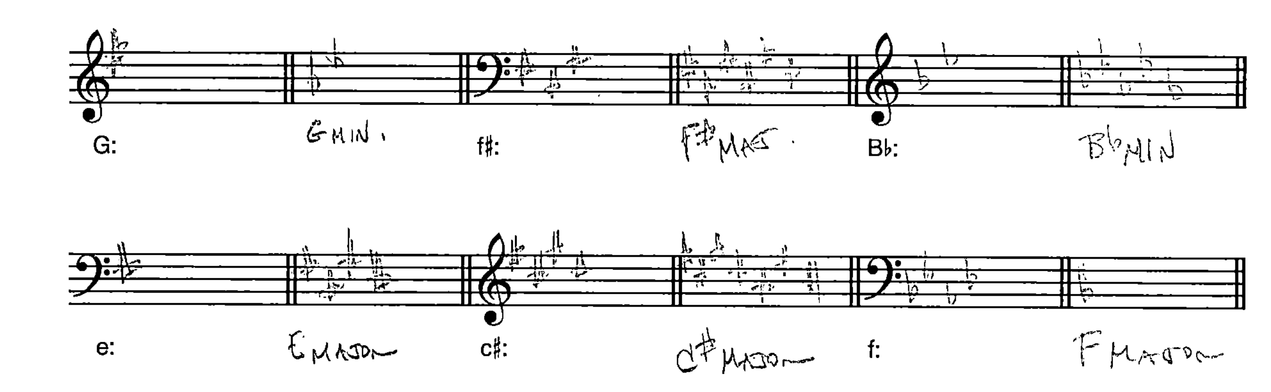

5 Here are a couple of brain teasers.

A. Multiply the number of accidentals in the key signature of G♭ major by the number of common types of minor scale ___18___. Subtract the number of whole steps in a major scale ___13___. Add the number of sharp major keys ___20___, and divide by the number of accidentals in the key signature of the parallel minor of F major ___5___. Your answer should be 5!

B. Divide the number of half steps in an octave by the number of accidentals in the key signature of F♯ minor. ___4___ Then add the number of whole steps in a major scale ___9___ and divide by the number of divisions of a beat in compound meter. ___3___ Now subtract the number of tritones in a harmonic minor scale. Your answer should be ___1___.

NAME ______________________________

6 In the following carol phrases, write the scale degrees above the last two notes (in Exercise 6B, the last two different notes). Indicate the type of melodic cadence: conclusive, less conclusive, or inconclusive. The key is given for each tune.

7 Many writers have remarked on the striking similarity between Beethoven's "Ode to Joy" melody and the principal theme of the last movement of Brahms's Symphony No. 1 in C minor. Do a melodic analysis of the Brahms tune similar to those given in Chapter 3 of the text. What similarities or differences do you find between the two pieces? BOTH CONSIST OF TWO 4 MEASURE PHRASES; THE CADENCE OF THE FIRST MOVES FROM 3–2 AND THAT OF THE SECOND FROM 3-2-1

8 Compose two four-measure melodic phrases in the staves that are provided. For each, incorporate the following characteristics:

A. Treble clef, minor mode, simple triple meter, and an inconclusive cadence.

B. Bass clef, major mode, compound duple meter, and a conclusive cadence.

CHAPTER 4

Triads and Seventh Chords

1 The three passages below are skeleton versions of the actual music. Write the root, chord type, and inversion for each triad in the excerpts. Then, look for melodic or chordal patterns—that is, different ways in which the notes are grouped together melodically or harmonically. Bracket these patterns in the music.

A. Wagner: Siegfried (simplified)

B. George Crumb: *Makrokosmos* I (simplified)

C. Liszt: *A Faust Symphony* (simplified)

2 **A.** Use each given note to construct six different root-position triads, three major and three minor. Above the staff, indicate which chord member the given note represents, and then below the staff, identify each triad. The first three chords are finished for you.

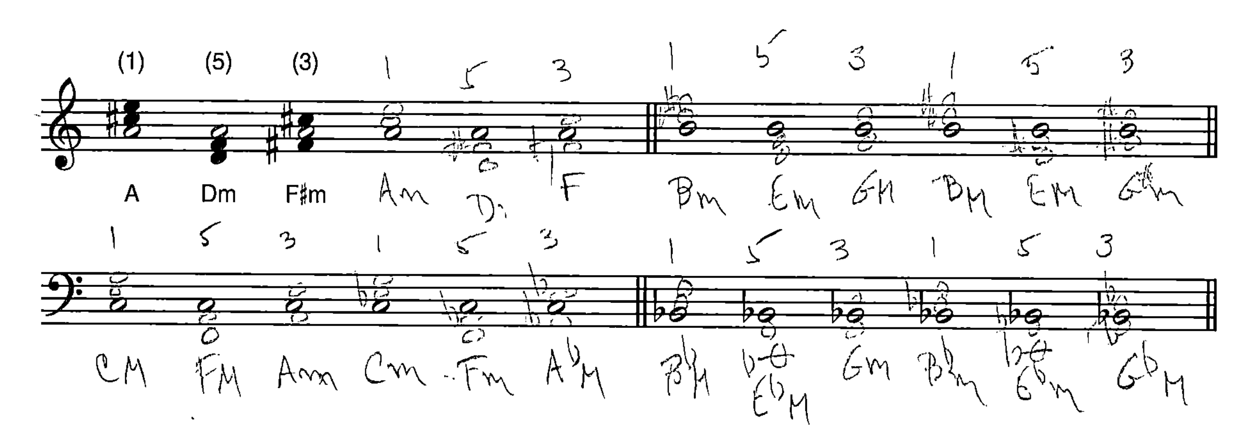

What two triads did you notate twice? ________ and ________. Excluding enharmonic spellings, what are the only two major or minor triads that do not appear? ________ and ________

B. Continue as above, but now construct only first-inversion triads; again, the first three chords are done for you.

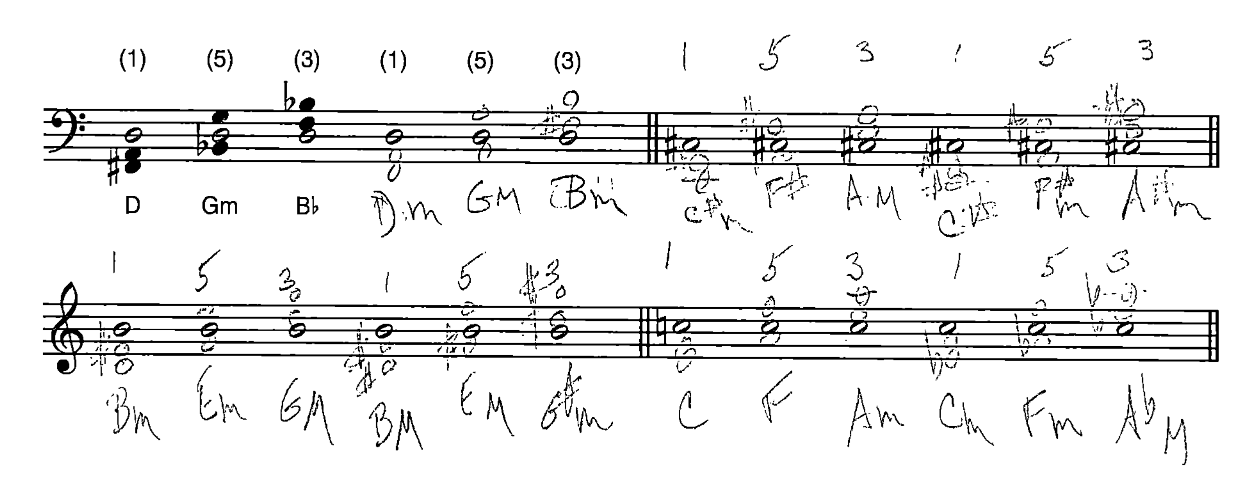

What two triads did you notate twice? ________ and ________. What are the only two major or minor triads that do not appear? ________ and ________.

3 Using the following pitches as the root, 3rd, 5th, or 7th (as specified above the staff) of the indicated chord types, write the chord *below* the given tone and give its inversion. The first three chords are done for you. Be sure to write the chord members consecutively from the top note down.

4 Using commercial chord symbols, analyze the following two progressions of four seventh chords. Then continue each pattern in strict fashion to the indicated final chords. To help you along the way, some of the chords or bass notes in each progression are provided. Make sure to distinguish between root-position chords and inversions.

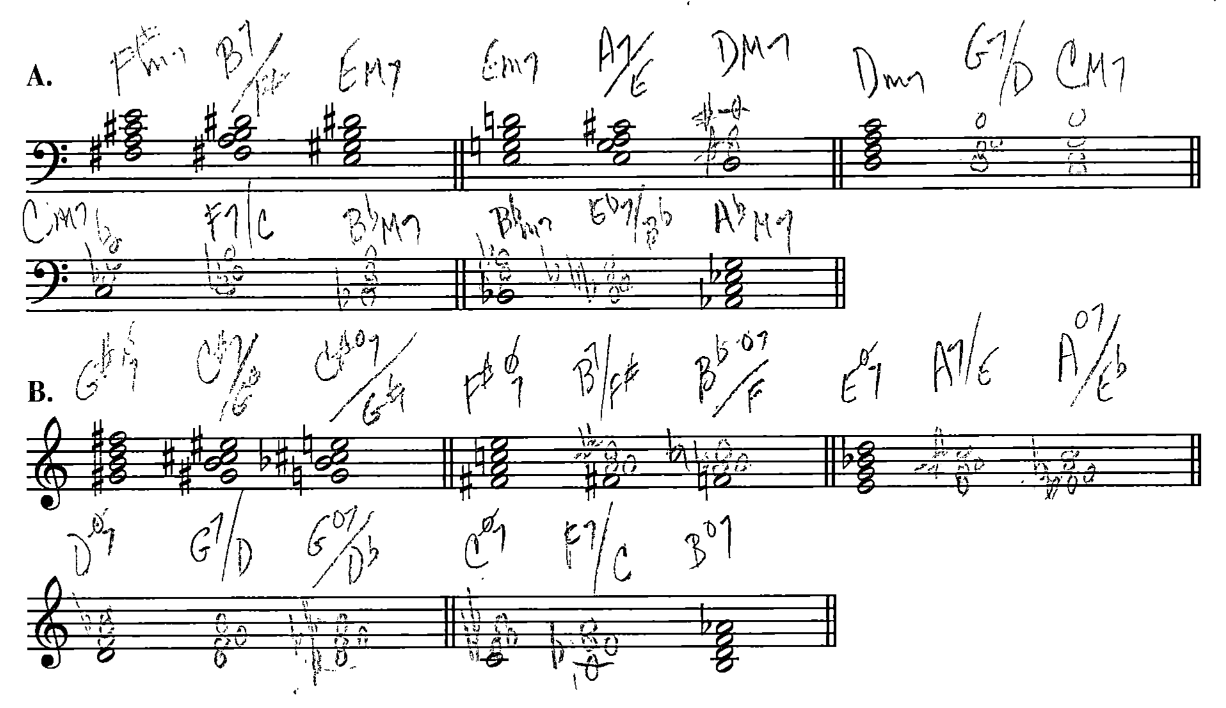

5 Using close structure *above* the lowest note (see the completed chords), realize the following figured-bass lines. Always take into consideration the key signature when spelling your chords, and be especially careful with altered intervals above the bass (♯, 6, ♭5, etc.).

NAME ______________________________

C H A P T E R 5

Musical Texture and Chordal Spacing

1 Indicate the basic types of texture found in Exercises 1A–1D: monophonic, chordal, melody with accompaniment, or contrapuntal. In excerpts with more than one type, indicate the measure(s) in which the change occurs.

A. MUSORGSKY: PROMENADE FROM *PICTURES AT AN EXHIBITION*

Texture: Monophonic 1+2 Chordal 2+3

B. DVOŘÁK: SYMPHONY NO. 9 IN E MINOR ("NEW WORLD"), II

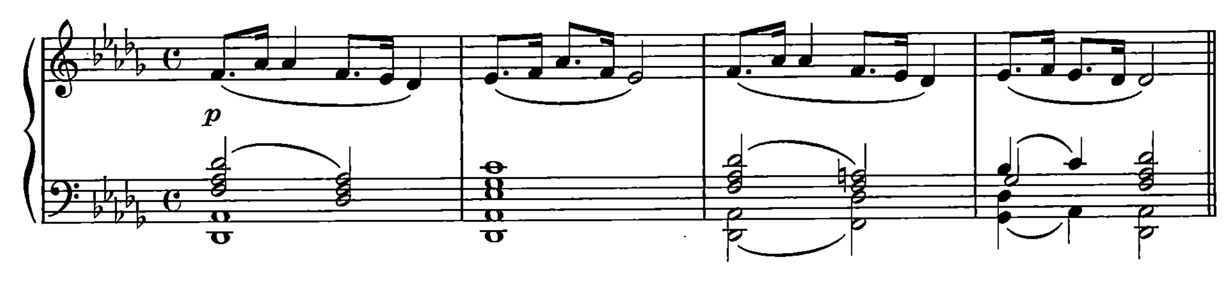

Texture: Melody with Accompaniment

C. BERLIOZ: "AMEN" FUGUE FROM THE DAMNATION OF FAUST
A - men, A -
A - men, A - - - - men, A -
A -
A - men, A -
- - men, A - - -
- men, A - - - men
- men, A - - - - - -
Texture: Contrapuntal (imitation)
D. BACH: "CHRIST, DER DU BIST DER HELLE TAG"
f
p
Texture: Chordal 1+2 Contrapuntal 3-6

2 The first section of the Mozart Minuetto utilizes a variety of textures within a comparatively few measures. Indicate the *successive* types of texture you find, using measure numbers.

MOZART: STRING QUARTET IN C MAJOR ("DISSONANT"), K.465, III

Texture: MONOPHONIC (1) CHORDAL (2-4) MONOPHONIC (4-6) CONTRAPUNTAL (6-11) MELODY WITH ACCOMPANIMENT (12-20)

3 Below each chord, supply the correct figured-bass number(s) and commercial chord symbol. Above each chord, indicate whether it is complete (C) or incomplete (I). How many seventh chords do you find? ___2___

CORELLI: TRIO SONATA IN A MAJOR, OP. 1, NO. 3, III (SLIGHTLY MODIFIED)

4 Analyze the Schumann four-part chorale setting by:

1. giving the figured-bass symbol below each chord
2. indicating the structure of each chord—C (close), O (open), or O/O (open/octave)—above the staff
3. marking doubled note(s) in each chord with vertical brackets within the staves.

The first three chords are done for you.

SCHUMANN: "FREUE DICH, O MEINE SEELE" FROM *ALBUM FOR THE YOUNG*, OP. 68

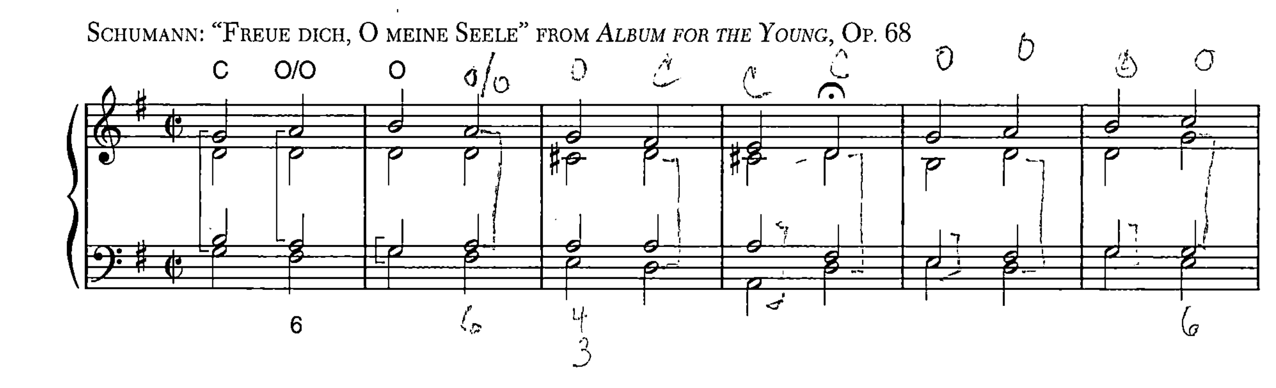

NAME ______________________

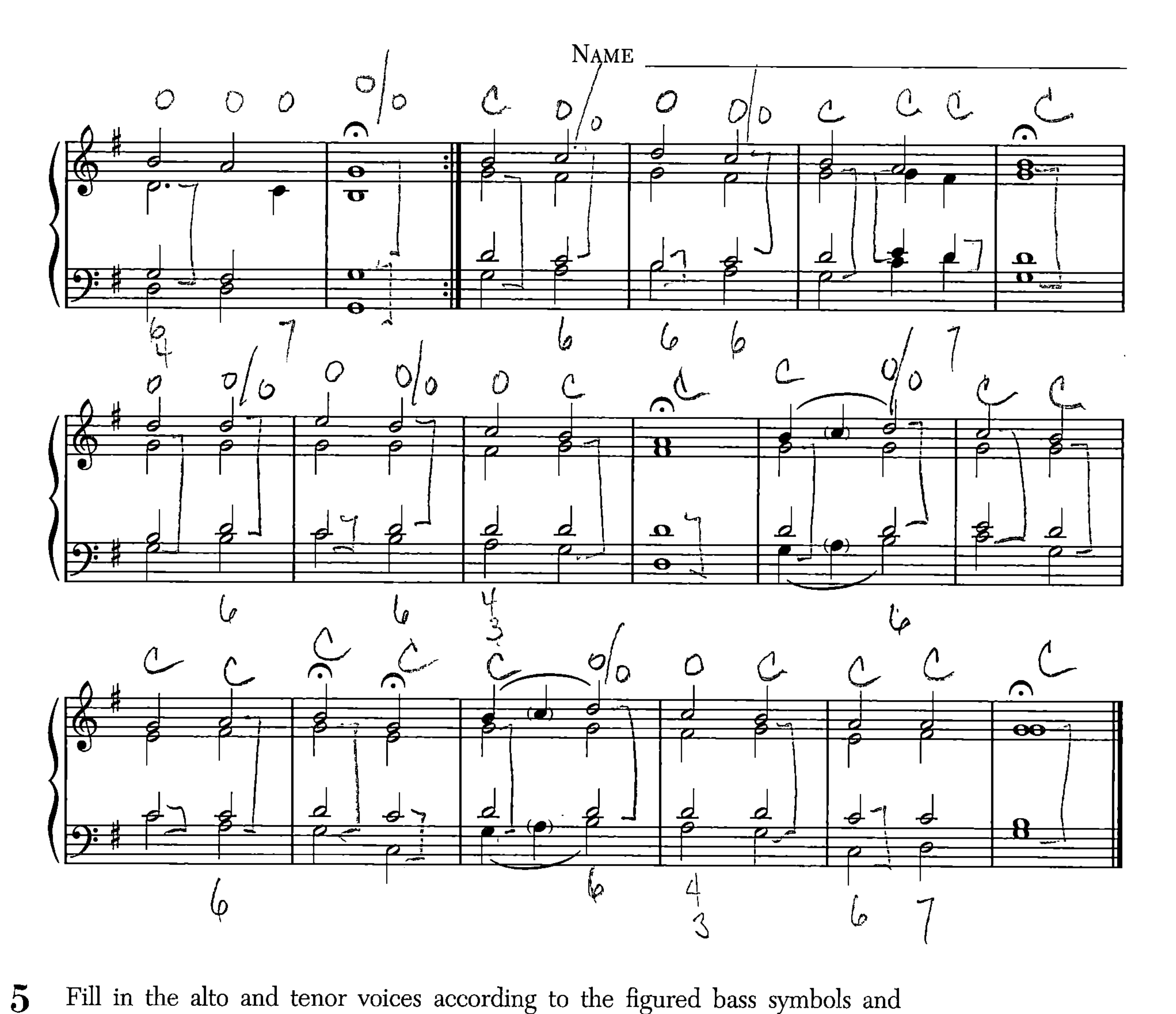

5 Fill in the alto and tenor voices according to the figured bass symbols and indications of chord structure, and identify each sonority with a commercial chord symbol below the staff. Be sure to observe the changing key signatures. Use the doubling procedures listed below unless otherwise instructed:

a. In root position, double the bass or root.
b. In first inversions of major and minor triads, double the soprano at the unison or octave.
c. In first inversions of diminished or augmented triads, double the chordal 3rd.
d. In seventh chords, avoid doubling any tones.

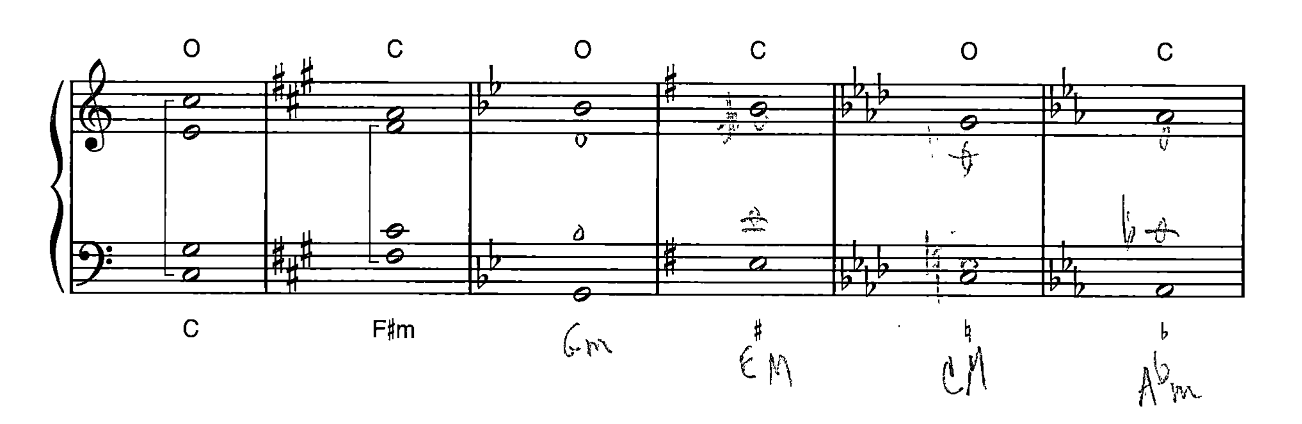

O/O
C
O
O/O
C
O
double the 5th
double the 3rd
triple the root
O
C
O
C
C
O
C
O
C
O
C
double the root

NAME ____________________

CHAPTER 6

Partwriting in Four-Voice Texture

1 These four-voice progressions contain various types of partwriting errors. Circle and identify the specific problem in each.

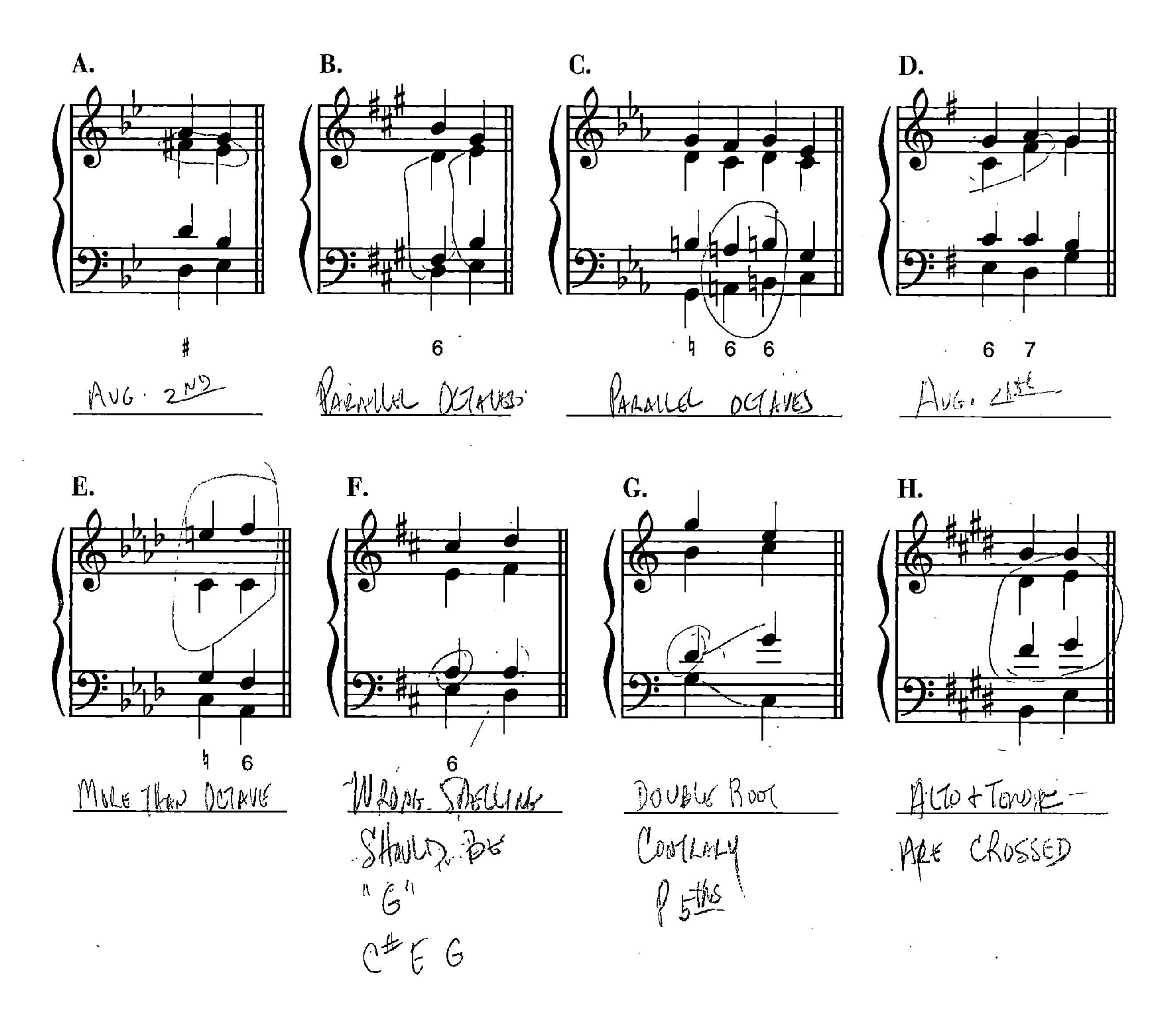

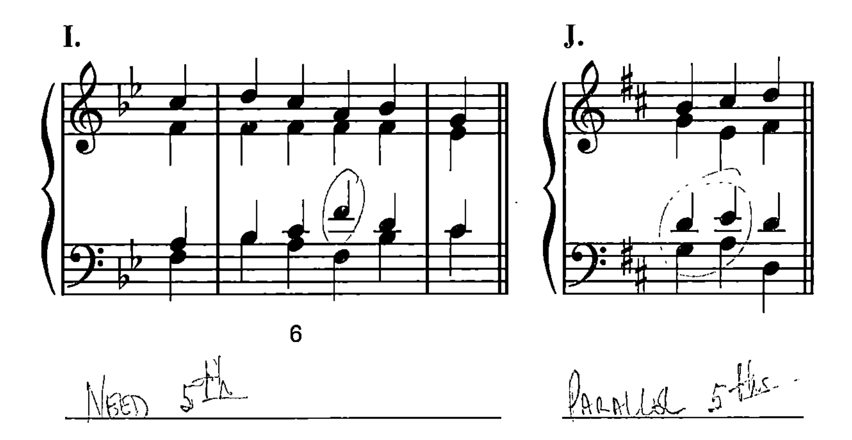

2 Using the chord structures indicated above the staff and the figured-bass numbers below the staff, fill in the alto and tenor voices to create a four-voice texture. Employ the doublings listed in Chapter 5, Exercise 5, unless otherwise instructed.

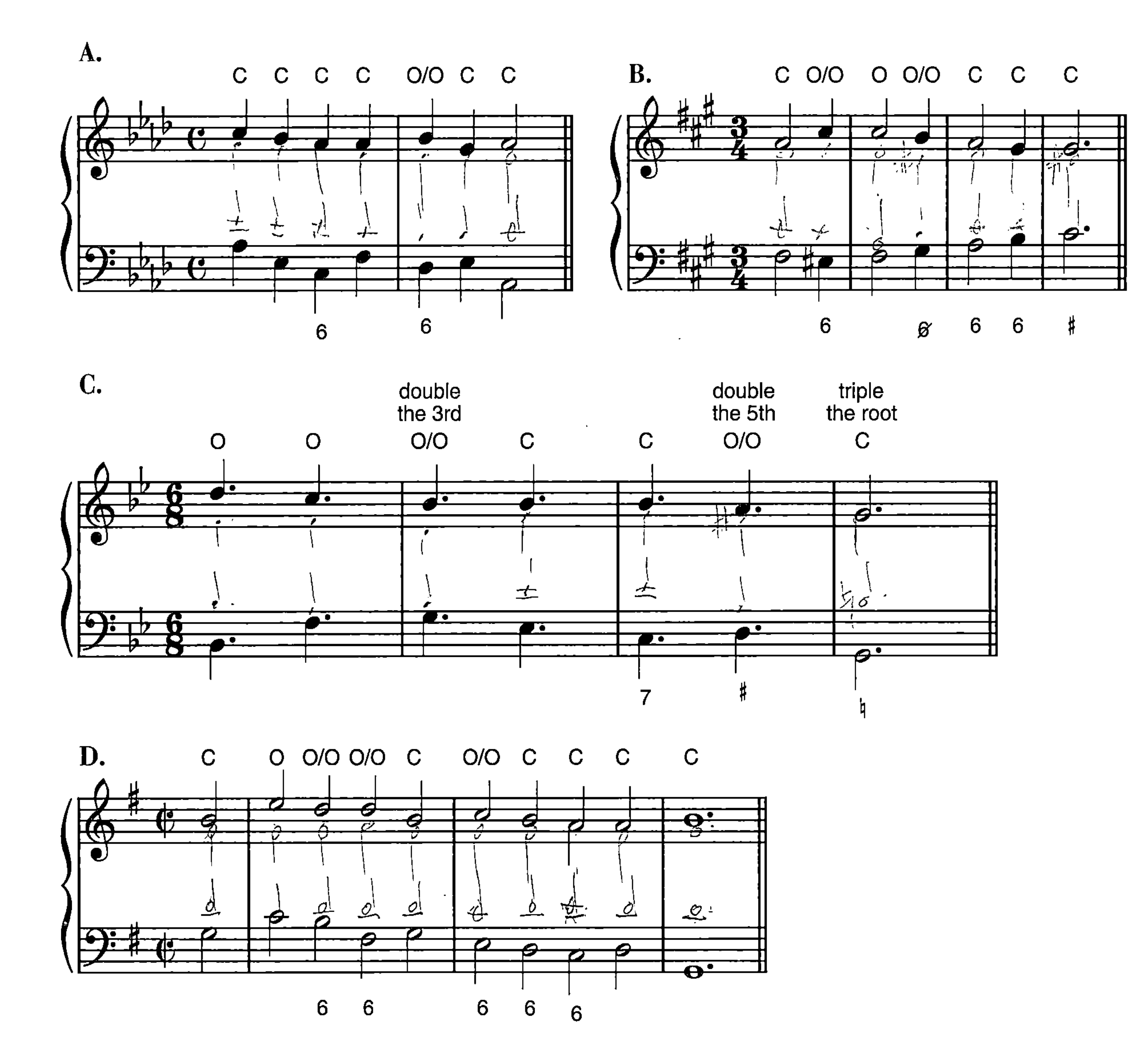

NAME ________________________________

C H A P T E R 7

Melodic Figuration and Dissonance I

NONHARMONIC TONES

1 The following excerpts contain various nonharmonic tones, which have been circled in the music. Above each one, identify its type, using the following abbreviations: P (for passing tone), N (neighboring tone), A (anticipation), IN (incomplete neighbor), S (suspension), and a prefix A (accented)—for instance AP = accented passing tone.

A. ANDREA GABRIELI: KYRIE FROM *MISSA BREVIS*

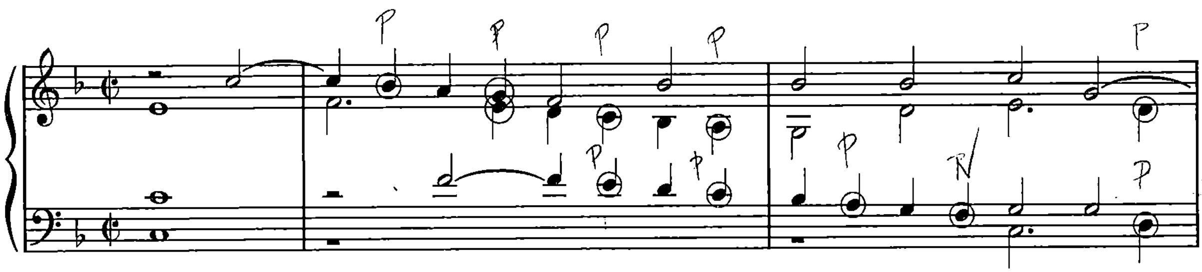

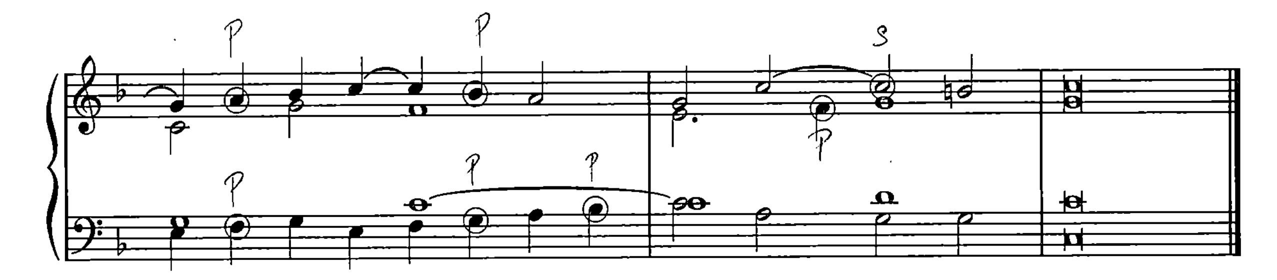

B. HAYDN: PIANO SONATA IN C MAJOR, HOB. XVI, NO. 35, III

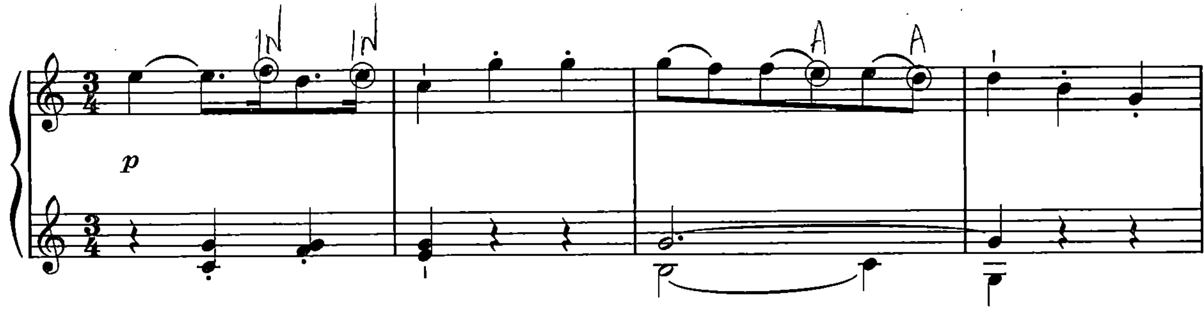

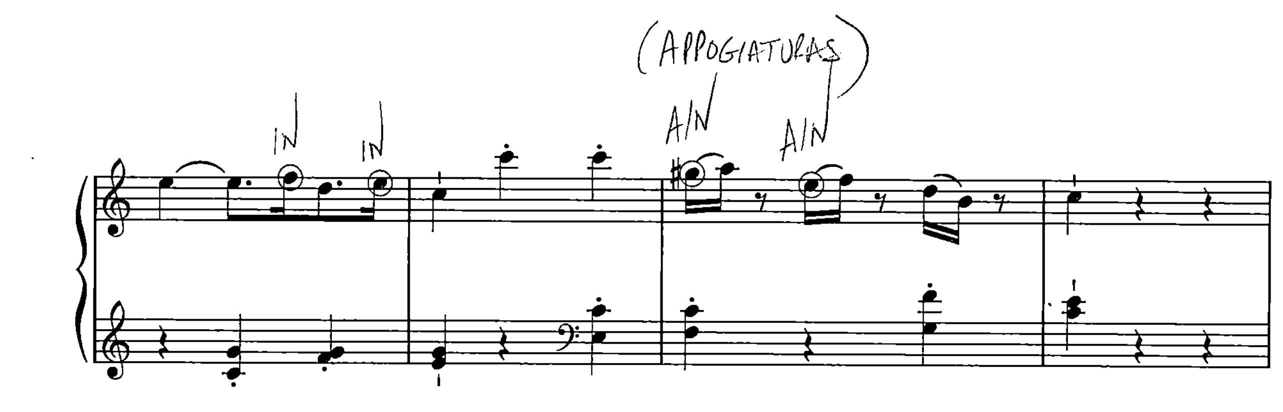

C. Bach: Fugue in E major from *Well-Tempered Clavier*, Book II

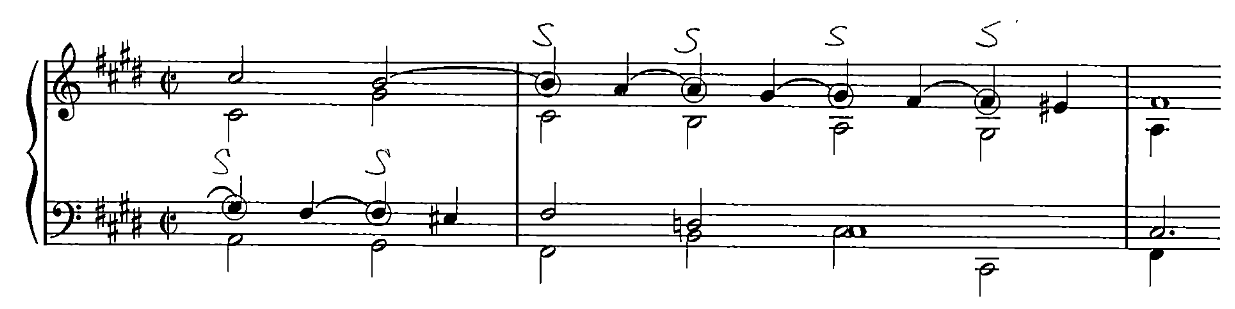

Do any of the excerpts contain melodic dissonances that fall into only *one* category, such as unaccented-stepwise, accented leaping, and so forth? If so, list them by letter in the space below, and indicate the specific category.

EX. A USES ONLY UNACCENTED - STEPWISE (P+N)

EX. C USES ONLY ACCENTED - STEPWISE (S)

2 Recopy the following piece, using the method of nonharmonic reduction found in Chapter 7 of the text. Use stemmed and unstemmed noteheads and connecting slurs, and identify each dissonant note. The first measure is done for you.

NAME ______________________________

3 Recopy the three passages below, adding appropriate nonharmonic tones in each passage as follows:

A. Only unaccented-stepwise tones (P, N, A) in any voice.

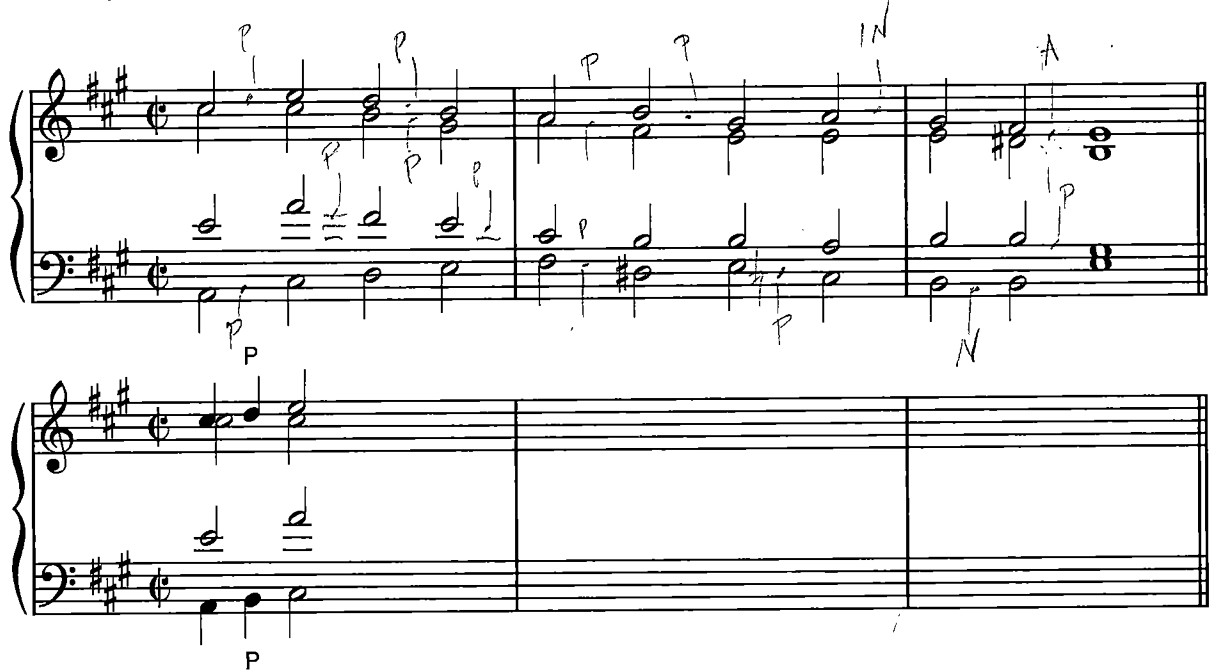

B. Only suspensions in the indicated voices.

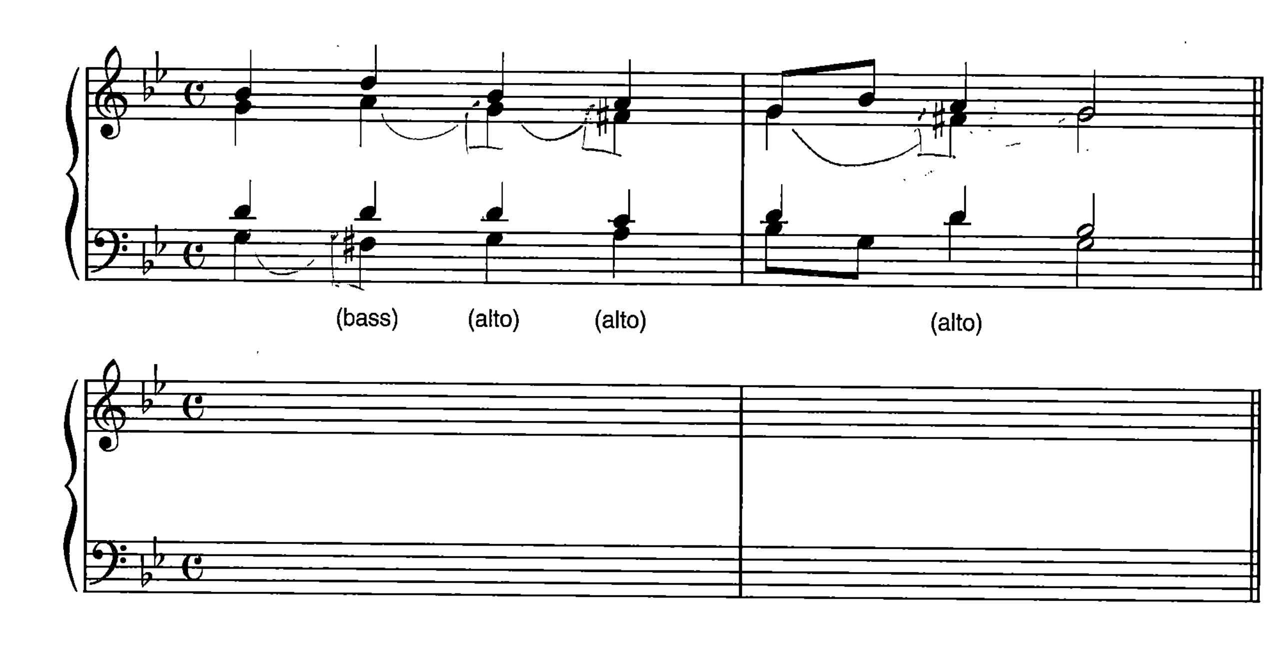

C. Only accented-stepwise tones (accented P or N) in any voice.

PART TWO

NAME ______________________________

CHAPTER 9

Tonic and Dominant Harmony

In those chapters that discuss specific chord functions (I, V, ii^{7}, etc.), you should first practice spelling the root position and inversions of these chords in various keys. For instance, spell I and V triads in the major keys of D, B♭, E, A♭, F♯, and B and i and V in the minor keys of F, C♯, A, E♭, and C.

When partwriting progressions, harmonizing melodies, or realizing figured-bass exercises, assume a four-voice texture unless instructed otherwise.

Many of the analytical exercises for these chapters include comments or questions for in-class discussion.

1 In the following cadential progressions, the key, roman numerals, soprano scale degrees, and chord structure are provided. Notate each progression (in four-voice texture), then identify the type of cadence, using one of the following abbreviations: PA (for perfect authentic), IA (imperfect authentic), and H (half).

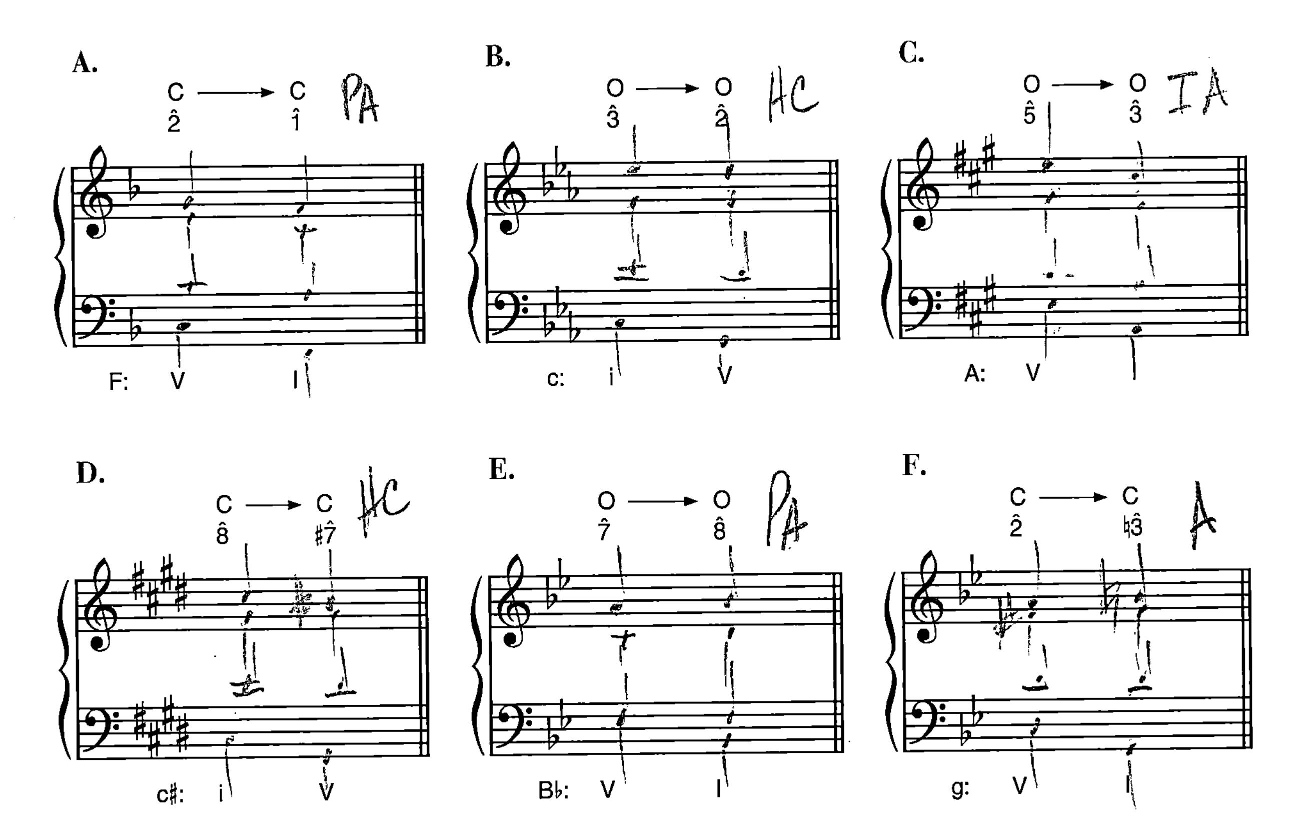

2 Indicate the key of each melodic phrase and the cadential scale degrees of the notes that are bracketed. Then write an appropriate cadence in four-voice texture, providing a roman-numeral analysis and indicating a cadence type; the choice of chord structure is up to you. Do not harmonize the unbracketed portions of each tune. In Exercise 2E, the melody is located in the bass part; you must choose an appropriate soprano line to fit it.

3 Realize the following figured-bass exercises (in four-voice texture) by filling in the alto and tenor. All nonharmonic tones are circled. The structure of the first chord is provided. In some exercises you can retain the same chord structure throughout, while in others you may have to change the structure periodically. Finally, provide a roman-numeral analysis; remember that the labels for all embellishing dominants should be enclosed in parentheses.

4 Set both melodic phrases for four voices, using only root-position I and V triads. First sketch in a plausible roman-numeral analysis and bass line, taking into account any passing or neighboring tones. Then add the remaining inner voices. Notice that in both places the rate of chord change is relatively slow. When the melody leaps to a different chord tone within the same harmony, you may either retain or change the chord structure.

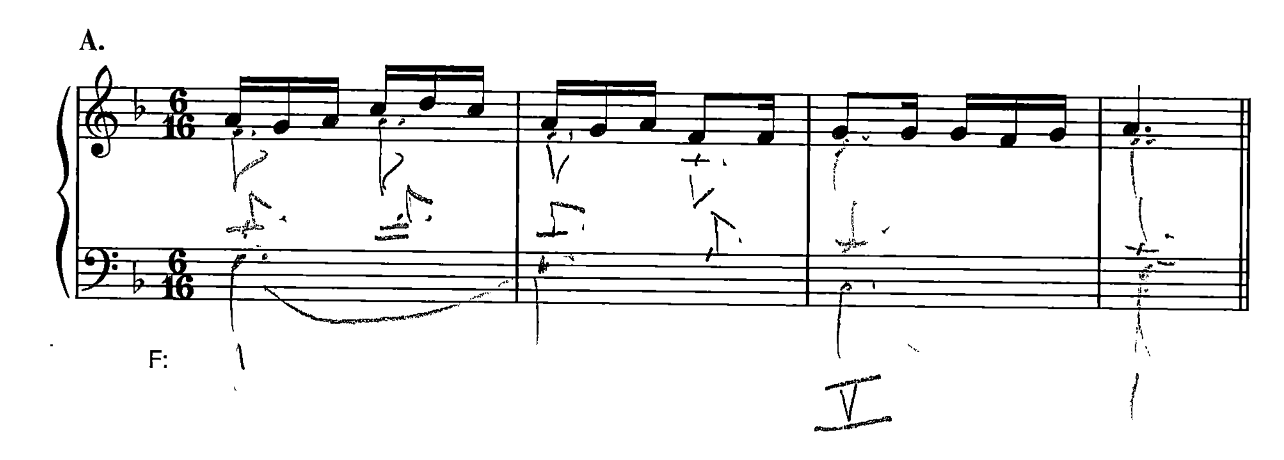

B.

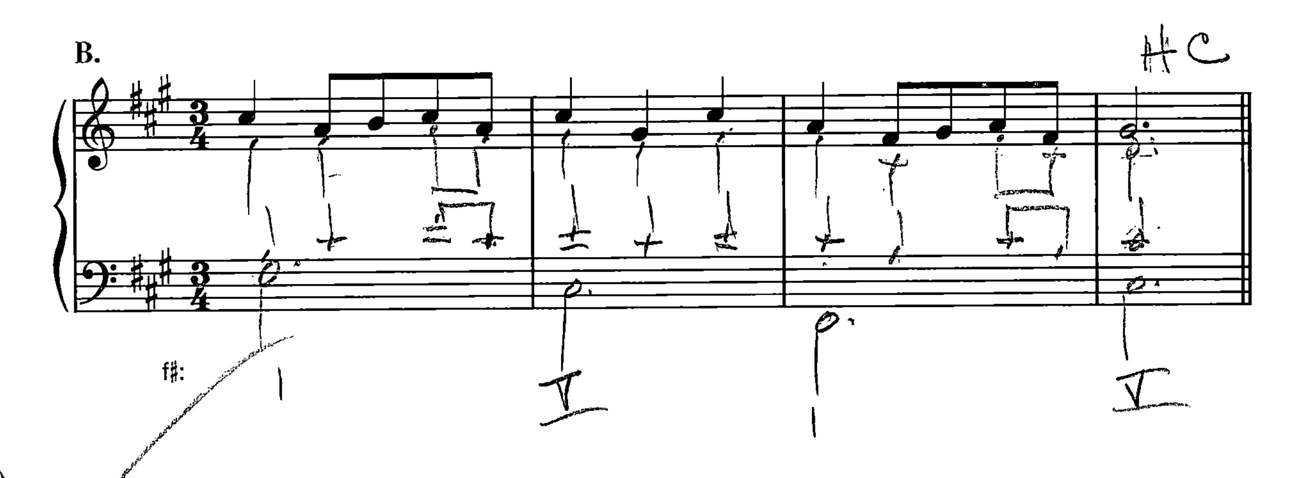

5 Provide each of the following excerpts with a voice-leading reduction (soprano and bass only) and a roman-numeral analysis. Be sure to distinguish between cadential/essential and embellishing dominant triads. Without the final A^5 in measure 3, what type of partwriting error would occur between the soprano and bass in Exercise 5C?

A. WEBER: MARCH OF THE PEASANTS FROM *DER FREISCHÜTZ*, ACT I

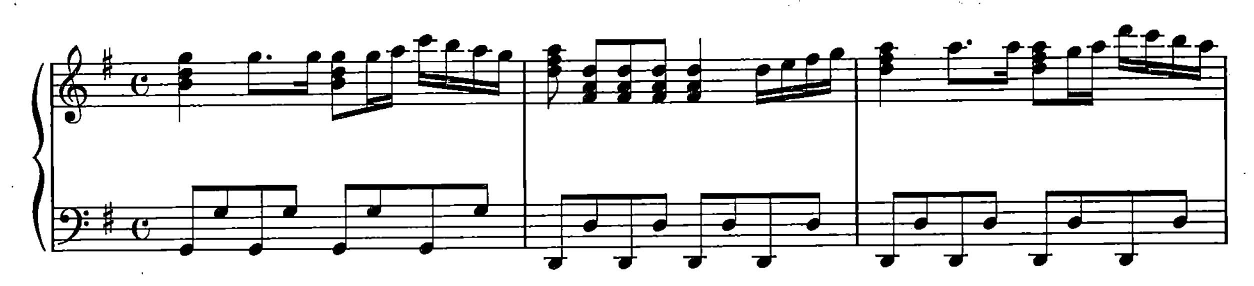

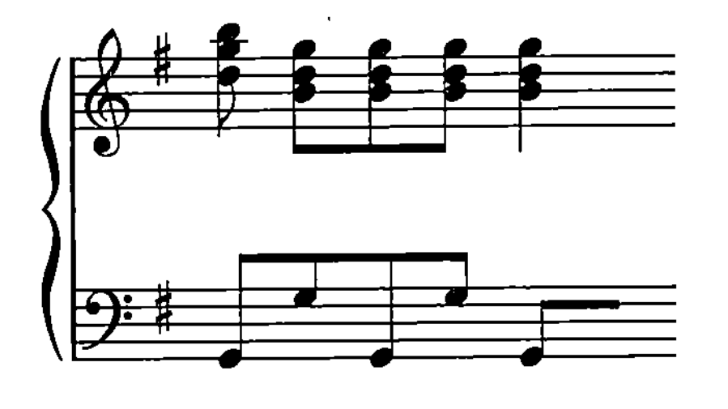

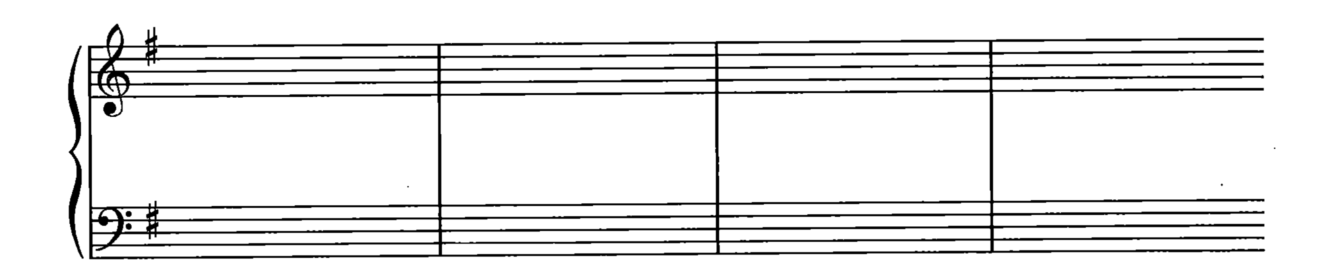

B. "Did You Ever See a Lassie?" (Scottish folk song)
C. Wagner: Overture to Der fliegende Holländer

NAME ______________________________

CHAPTER 10

Rhythm and Meter II

RHYTHMIC DEVIATIONS AND METRICAL DISSONANCE

1 Write some original melodies (only a few measures each) that illustrate the following rhythmic devices:

a. hemiola
b. superimposed meter
c. syncopation at the subbeat level
d. implied meter change within a single meter signature

2 Brahms's Symphony No. 2 in D major is a virtual compendium of rhythmic and metric deviations. Below each of the following excerpts, identify the most prominent device.

A. III, MM. 1–4, 33–36

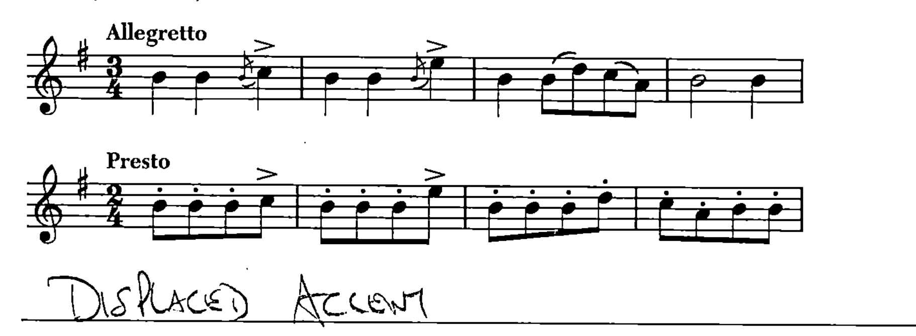

B. I, MM. 238–42

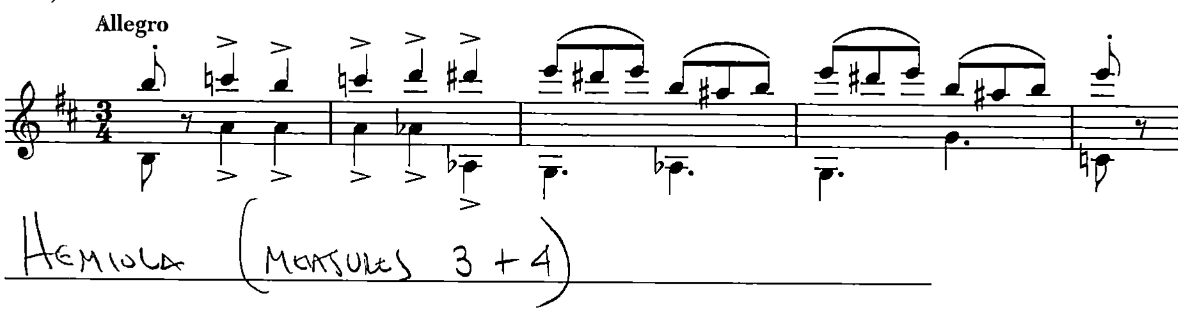

C. I, MM. 118–22 (NOTE POSITIONING OF OCTAVE LEAPS)

D. IV, MM. 312–17

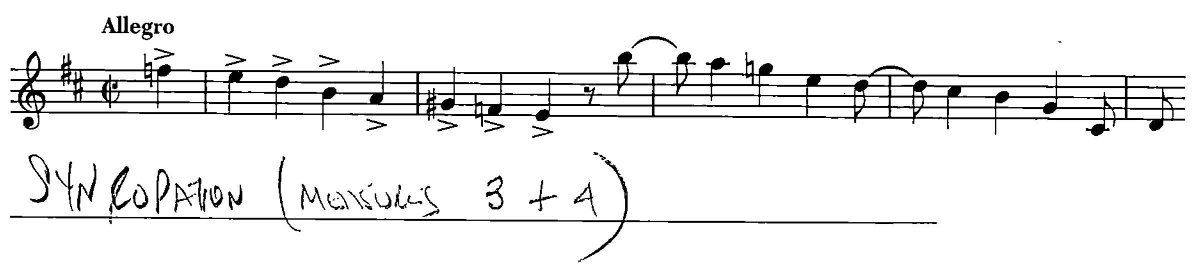

E. I, MM. 77–81

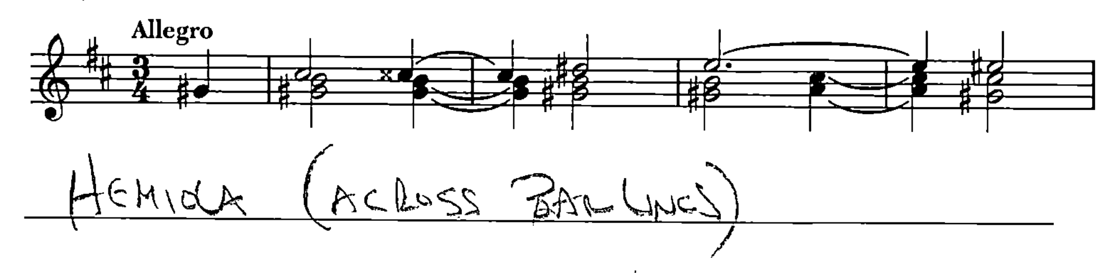

MELODIC GROUPINGS IN MM 1–5 IMPLY 5/2

HEMIOLA MEASURE 6 THRU 8

3 The rhythmic/metric aspect of this short passage has long fascinated music scholars. Try your hand at explaining what is "going on" in the portion that is bracketed. Be ready to defend your opinion in class discussion.

Mozart: Piano Quartet in G minor, K.478, I

NAME ____________________

CHAPTER 11

The V^7 and I^6 Chords

1 An upper voice is provided for each of the following brief progressions; the stemming of the given part indicates whether it is the soprano, alto, or tenor. Write out the bass line that is indicated by the roman numerals. Then fill in the remaining voices, using correct partwriting. You may need to omit the chordal 5th of the V^7 in some instances to prevent parallel 5ths.

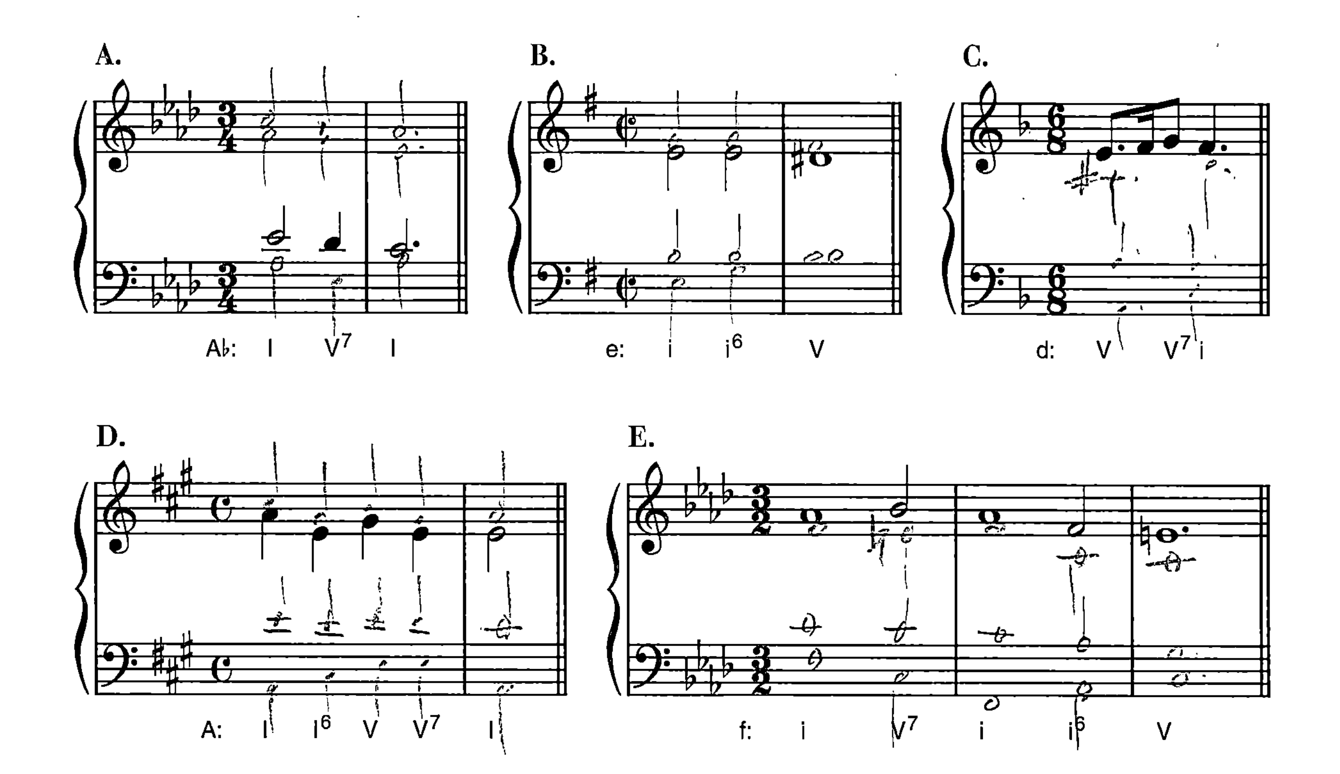

2 The following three progressions consist of a bass line and an indicated V^7. For each, write out the dominant chord in four voices, and then use the chords that precede and follow it to approach and resolve the chordal 7th. The letters above the staff indicate where the 7th acts as a passing tone (P), neighboring tone (N), or incomplete neighboring tone (IN). Add the missing roman numerals.

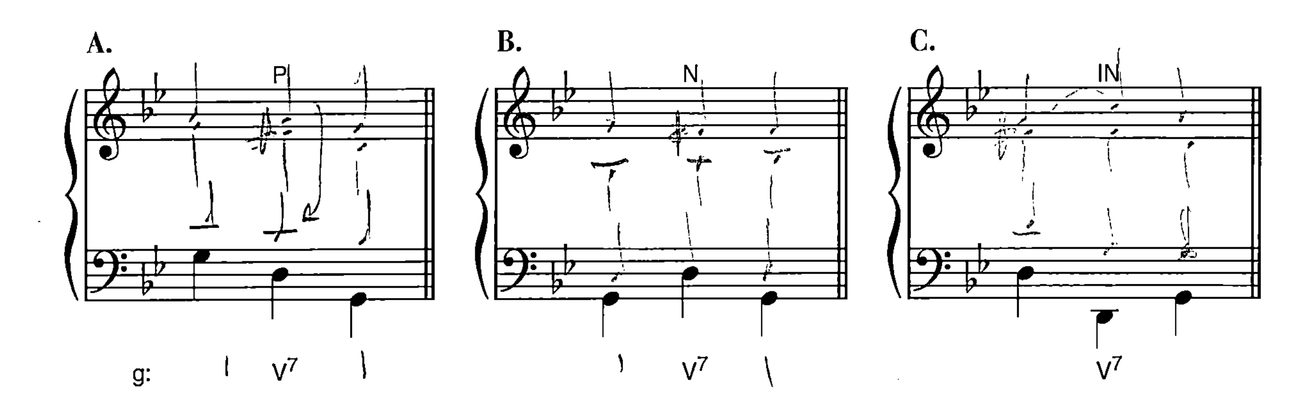

3 Realize the following figured-bass exercises, and provide a roman-numeral analysis. Enclose all embellishing V's and V^7's in parentheses. Be able to explain the treatment of the chordal 7th in the various V^7's. Exercises 3B and 3D include some nonharmonic tones.

Name ______________________________

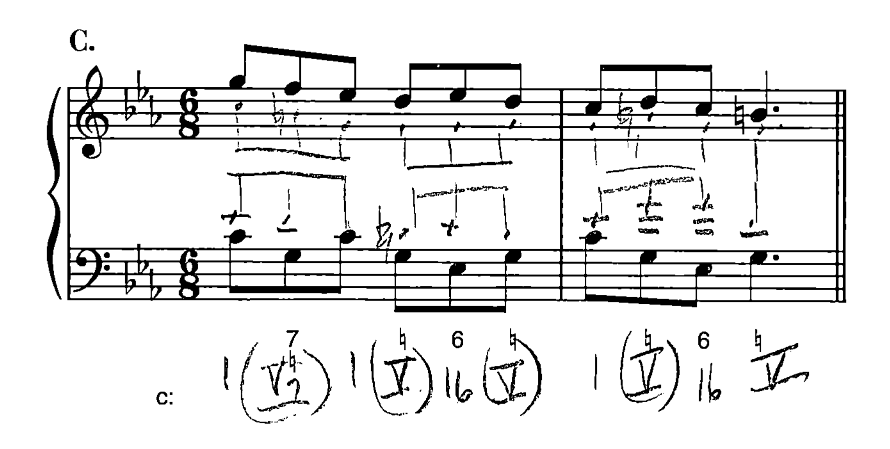

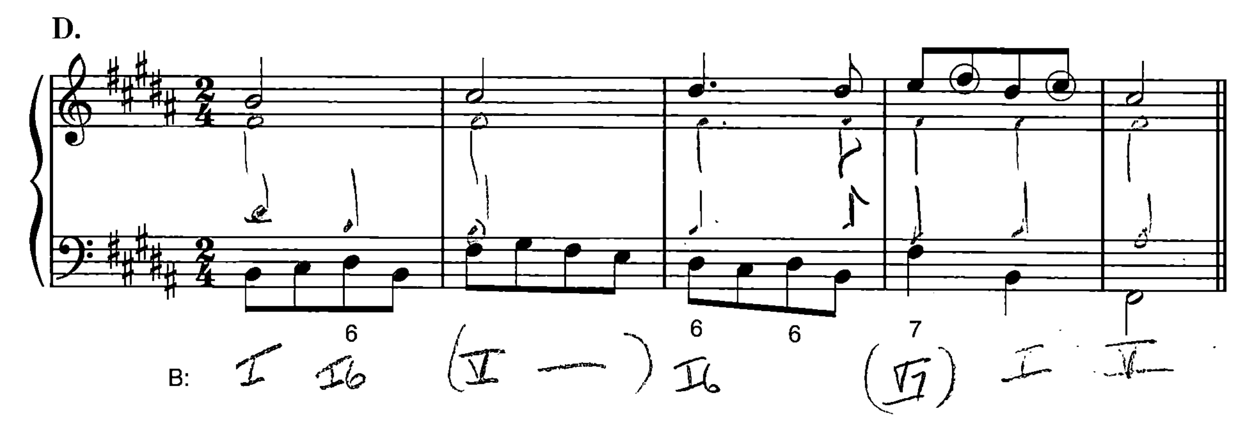

4 Harmonize the following two melodic lines, using *only* I, I^6, V, and V^7 chords. You may wish to elaborate your bass line and inner voices with some passing or neighboring tones, but be careful not to inadvertently create parallel 5ths.

5 For each of the following excerpts, make a two-voice (soprano and bass) voice-leading reduction, adding roman numerals below the staff.

A. In your reduction, be sure that you do not include unessential melodic notes such as passing tones. What harmonic device does this passage exemplify? Where does the F^5 resolve?

BEETHOVEN: BAGATELLE IN C MAJOR, OP. 119, NO. 8

B. Locate the chordal 7th of the V^7 and its resolution in this passage. How is it derived? Be sure to examine the entire soprano line before answering.

BACH: BRANDENBURG CONCERTO NO. 4 IN G MAJOR, III (SIMPLIFIED)

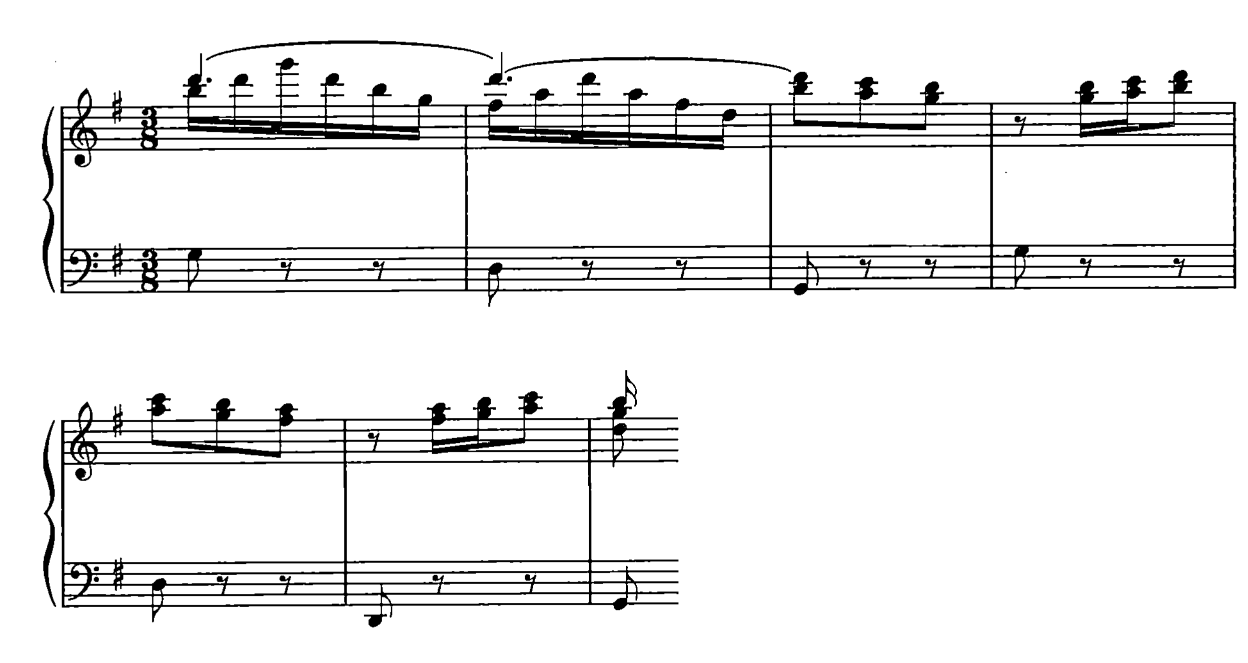

NAME ______________________________

C. Each of the following two excerpts quotes the first eight measures of the Trio section of a Minuet. In what ways are these excerpts similar? Despite these similarities, how does the underlying voice leading of the upper voices differ? *Hints:* In the Haydn excerpt, what aspect of the notation offers a clue to the voice leading in the soprano? In the Mozart excerpt, what notes lead to the important notes of the soprano line at the beginning of measures 3 and 7?

HAYDN: SYMPHONY NO. 97 IN C MAJOR, III, TRIO

Mozart: Symphony No. 39 in E♭ major, III, Trio (simplified)

NAME ____________________

C H A P T E R 12

Phrase Structure and Grouping

1 Examine the melodic passages below and follow the instructions that accompany each one.

A. Supply scale degrees above the cadential notes, and then write the harmonies that are implied by them below the staff. How many phrases are there in this excerpt, and what is the tonal relation between them?

"WHEN THE SAINTS GO MARCHING IN" (AFRICAN-AMERICAN SONG)

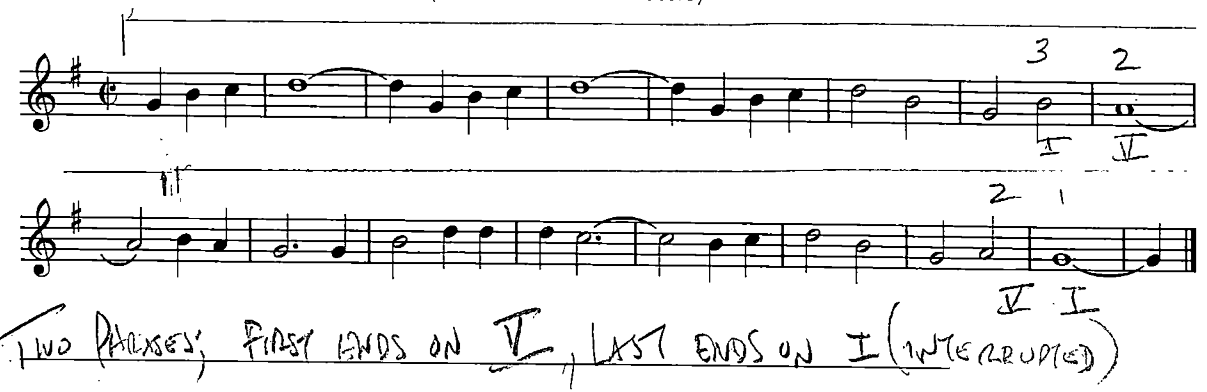

B. Repeat the procedure given above. What cadential similarities do you observe between this tune and the preceding one? What curious inconsistency exists between its title and meter?

"WALTZING MATILDA" (AUSTRALIAN BUSH SONG), CHORUS ONLY

BOTH HAVE TWO PHRASES AND SAME CADENCES. A "WALTZ" IS IN 3/4, NOT 4/4

C. This passage consists of a thematic statement by the soloist and the orchestral tutti. Use the procedure given above, and identify the overall form of this passage. What do we call the melodic design in the first eight measures?

MOZART: CONCERTO FOR HORN AND ORCHESTRA IN E♭ MAJOR, K.447, III

D. Name two ways in which this passage deviates from the norm. Focusing on the change in dynamics and thematic content will help you determine what these deviations are.

HAYDN: STRING QUARTET IN B♭ MAJOR, OP. 71, NO. 1, I

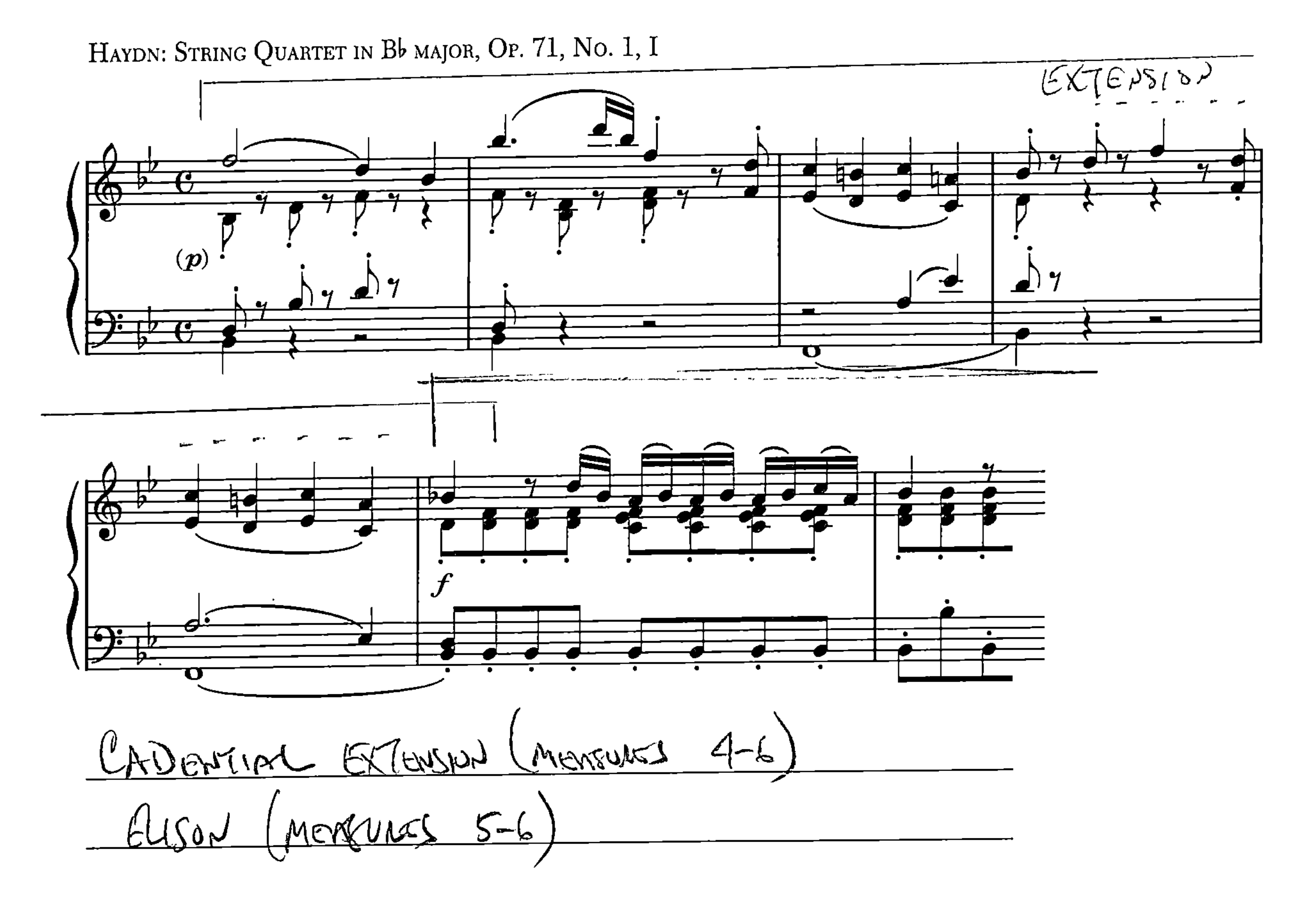

NAME ______________________________

E. How is each phrase divided? What is the term used to describe each segment?

SCHUBERT: "VIOLA," OP. 123 (TEXT OMITTED)

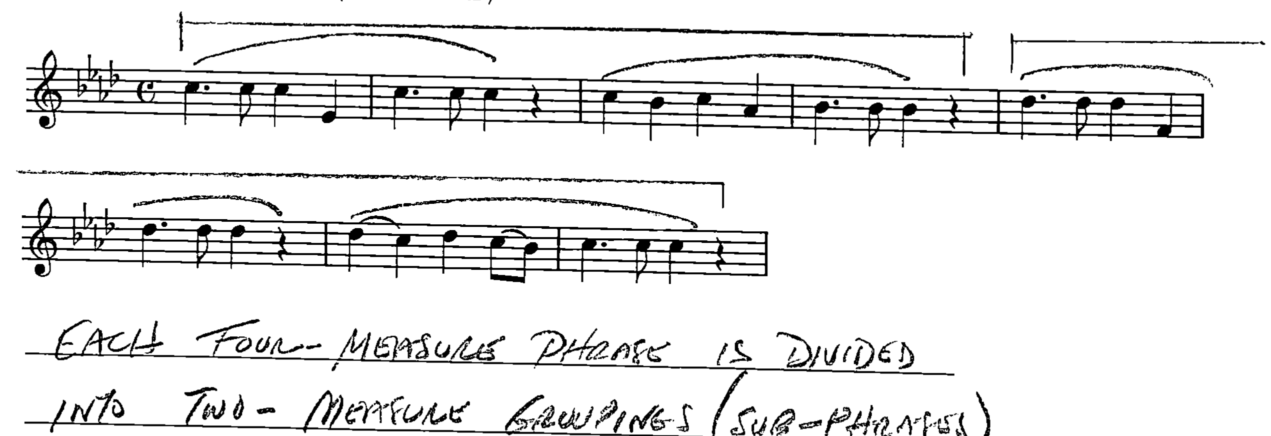

EACH FOUR-MEASURE PHRASE IS DIVIDED INTO TWO-MEASURE GROUPINGS (SUB-PHRASES)

F. Compare this pair of melodic statements. What technique has Brahms employed in the second one?

BRAHMS: SYMPHONY NO. 4 IN E MINOR, I, MM. 1–4 AND 246–58

RHYTHMIC AUGMENTATION

G. What happens to the melody in each group of two successive slurs? What change does this bring about in the meter?

STRAVINSKY: *LE SACRE DU PRINTEMPS*

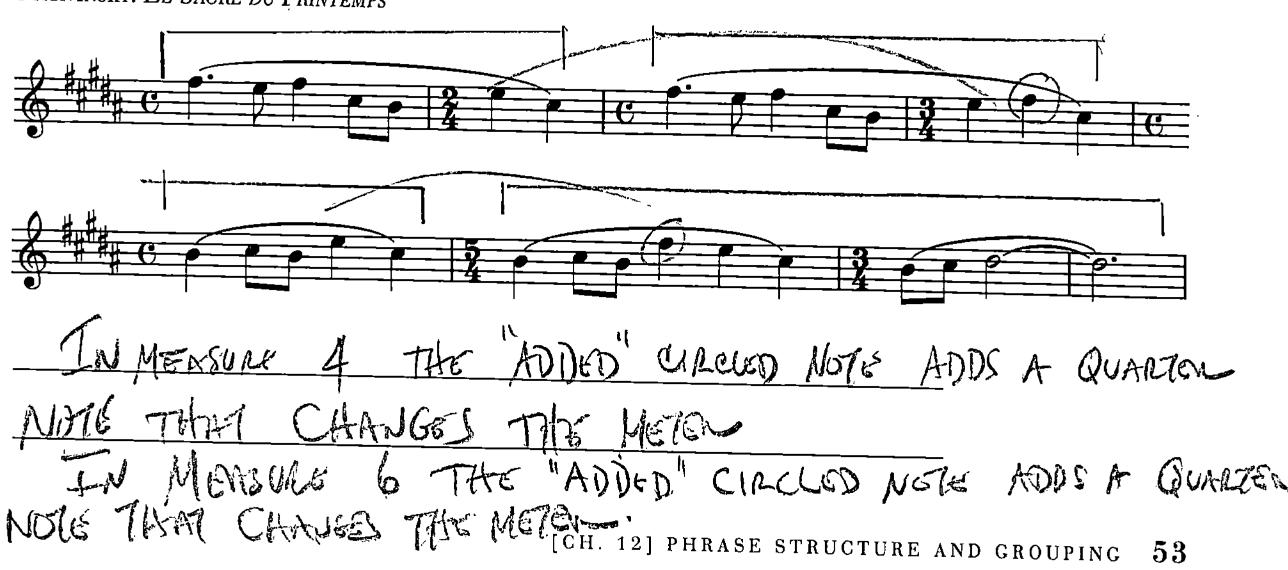

IN MEASURE 4 THE "ADDED" CIRCLED NOTE ADDS A QUARTER NOTE THAT CHANGES THE METER

IN MEASURE 6 THE "ADDED" CIRCLED NOTE ADDS A QUARTER NOTE THAT CHANGES THE METER.

2 Compose an original melodic passage that incorporates one of the following devices:

a. two parallel phrases
b. phrase elision
c. cadential extension

3 J. S. Bach originally wrote his Two- and Three-Part Inventions as lessons in keyboard technique and compositional organization for his son Wilhelm Friedemann. They are superlative examples of how short motivic ideas can be expanded, inverted, or developed in other ways into complete pieces. The familiar C-major Two-Part Invention given below is notable for its great economy of means. The opening melodic statement, on which the piece is based, is made up of two separate motives: a stepwise motion that rises a 4th (marked *x*), and two falling 3rds (marked *y*). Trace these motives throughout the invention, paying attention to the various ways in which they are developed; you may write on the score. Compare measures 3–4 with measures 11–12 and with measures 19–20; what happens to the motives in these measures?

BACH: TWO-PART INVENTION IN C MAJOR

NAME ____________________

4 This excerpt, given in simplified form, contains a principal melody and a motivic idea that assumes a secondary or accompanimental role in the texture. Identify this motive and circle all its appearances, noting any examples of melodic inversion or mirroring.

Brahms: Symphony No. 1 in C minor, IV

NAME ______________________

CHAPTER 13

Linear Dominant Chords

V^6, VII$^{\circ 6}$, AND INVERSIONS OF V^7

1 In each of the following, an embellishing dominant chord is indicated in a specific key. Complete the upper voices of the chord in four-part texture. Then precede and follow it with tonic harmony (I or I^6). In the case of inverted V^7's, be sure that the chordal 7th is approached and resolved correctly; you may wish to refer to the models in the text. Provide each of your tonic chords with roman numerals.

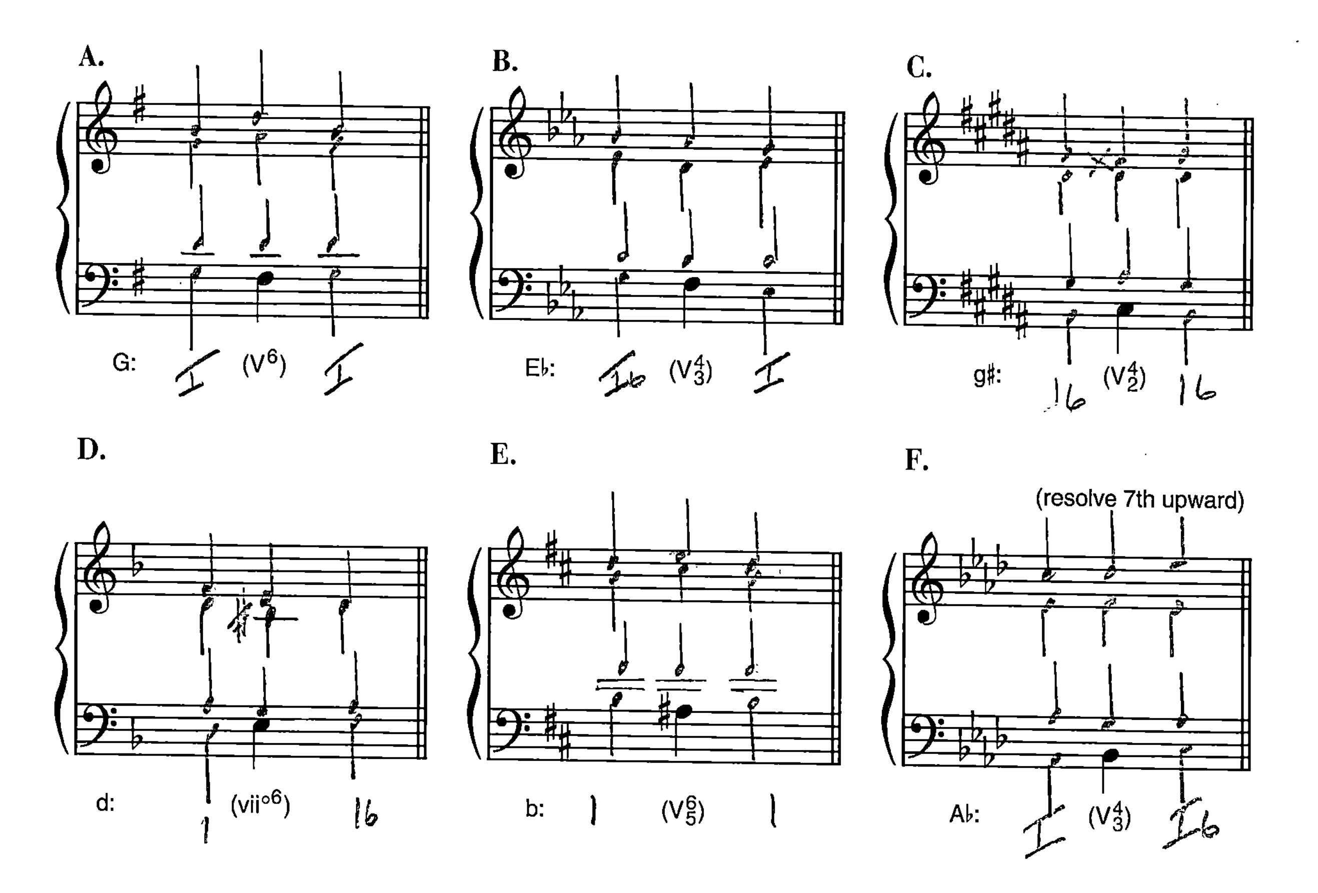

2 Realize the following figured-bass exercises, and supply a roman-numeral analysis. Enclose any embellishing dominants in parentheses.

Exercise 2D omits the symbols customarily found below the bass. In such an instance of *unfigured bass,* you must deduce the implied harmonies from the outer voices. Other exercises with an unfigured bass appear in subsequent chapters.

D.

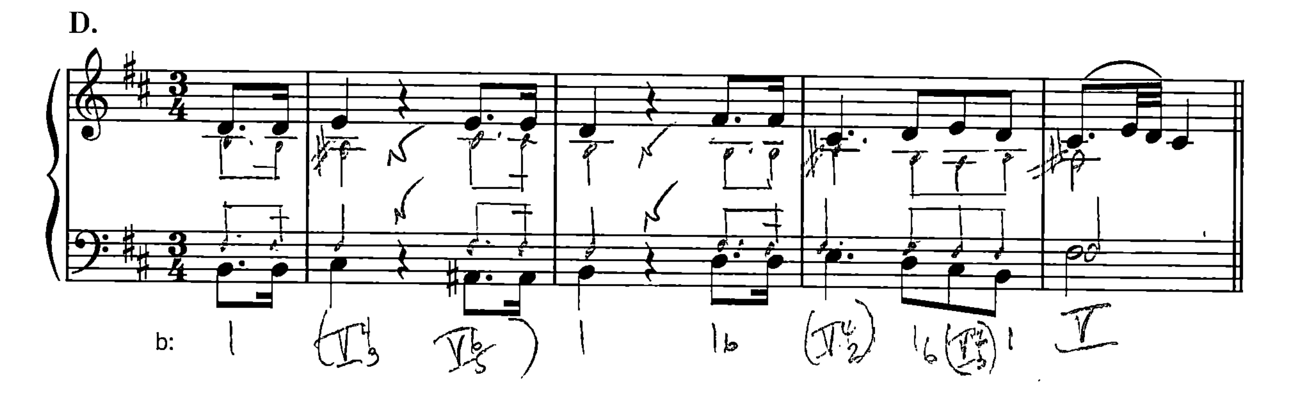

E.

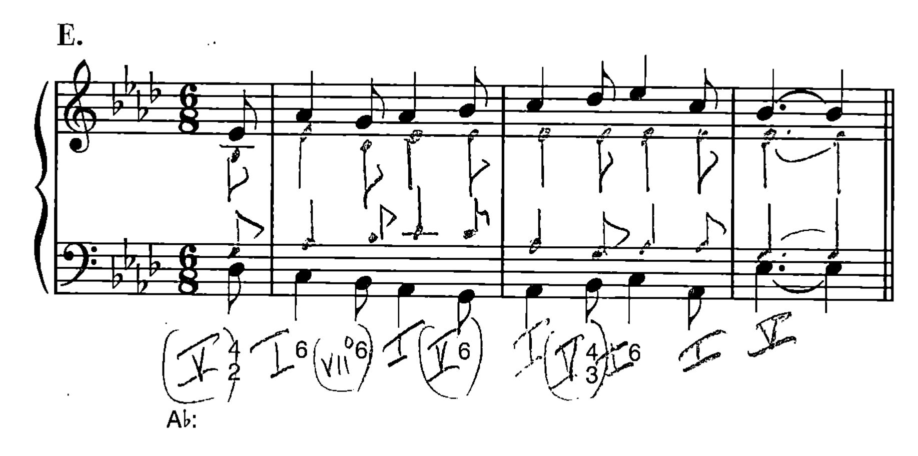

3 Indicate which of the following excerpts illustrate an instance of displaced 7th, prolonged dominant, delayed resolution, or resolution of the chordal 7th in a different voice. Mark your answers on the scores.

A. BEETHOVEN: PIANO SONATA IN E♭ MAJOR, OP. 81A ("LES ADIEUX"), I

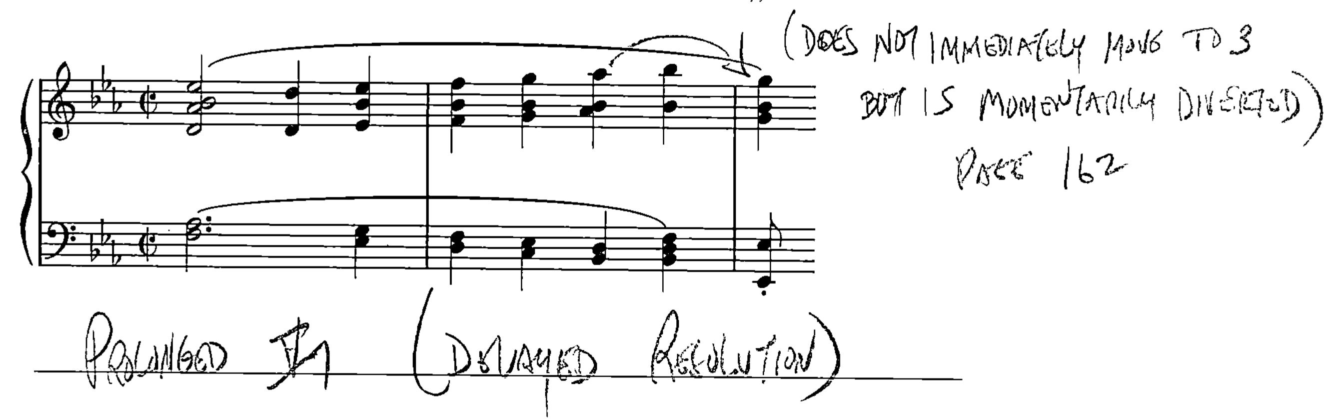

B. Mozart: Symphony No. 41 in C major ("Jupiter"), II

C. Haydn: Piano Sonata in C major, Hob. XVI, No. 21, II

DOES NOT IMMEDIATELY RESOLVE TO TONIC

DELAYED RESOLUTION (PROLONGED DOMINANT)

D. Bach: Recitative from Cantata No. 80, *Ein' feste Burg ist unser Gott*

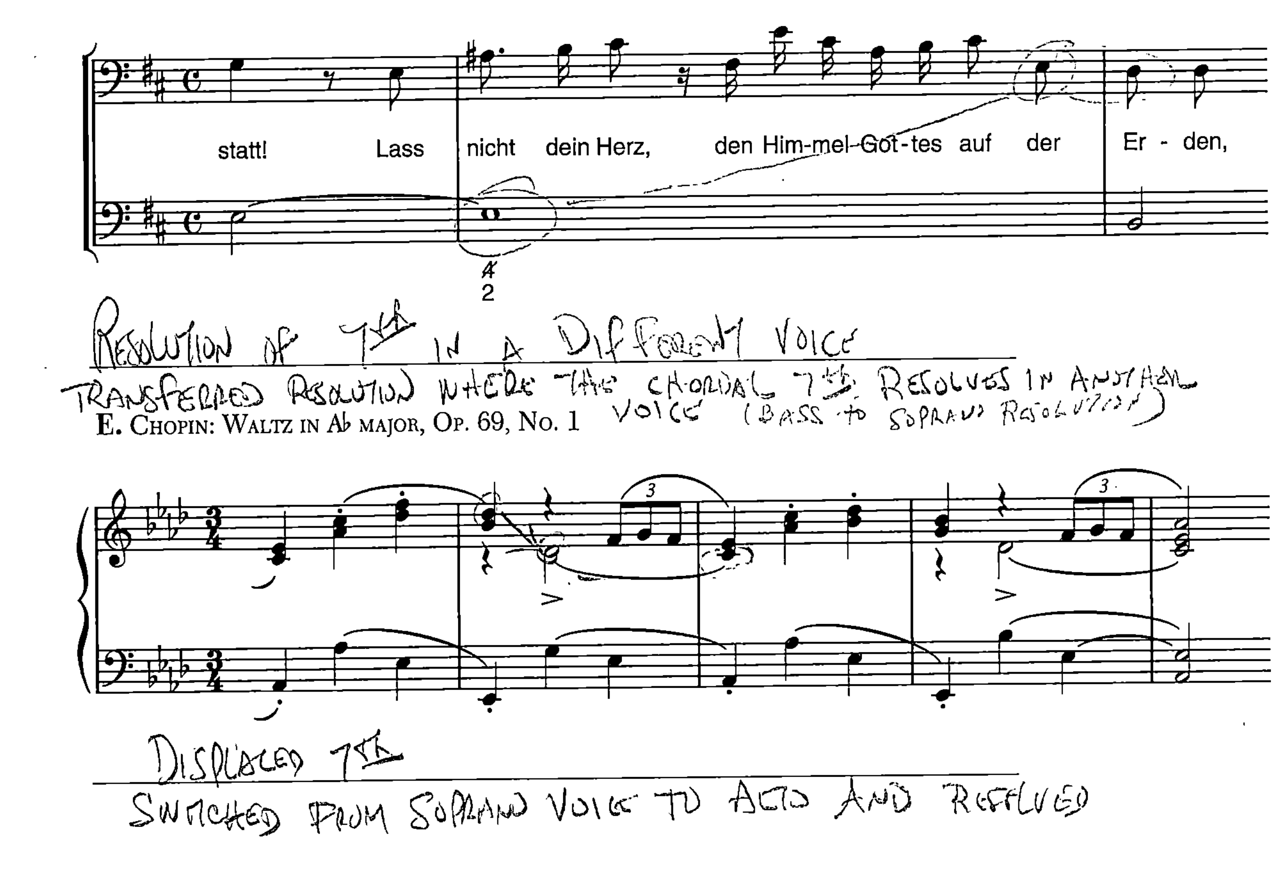

E. Chopin: Waltz in A♭ major, Op. 69, No. 1

4 Harmonize two of the following three melodies, working out an interesting bass line by employing various inversions or embellishing dominants. Then provide a roman-numeral analysis of your melody harmonization. In the first tune, delay the cadence until the final two bars. Exercise 4B is an elaborated version of a tune from Chapter 10; write a different bass line for it here.

A. "GLASGOW" (HYMN TUNE)

F:

B.

b:

C.

E:

5 The piano sonatas of Beethoven contain a number of opening statements that utilize only tonic and dominant-family harmonies. Three of these are quoted below. Make a voice-leading reduction and provide a roman-numeral analysis for each.

A. How is the tonic harmony prolonged in this example?

BEETHOVEN: PIANO SONATA IN C MINOR, OP. 13 ("PATHÉTIQUE"), II

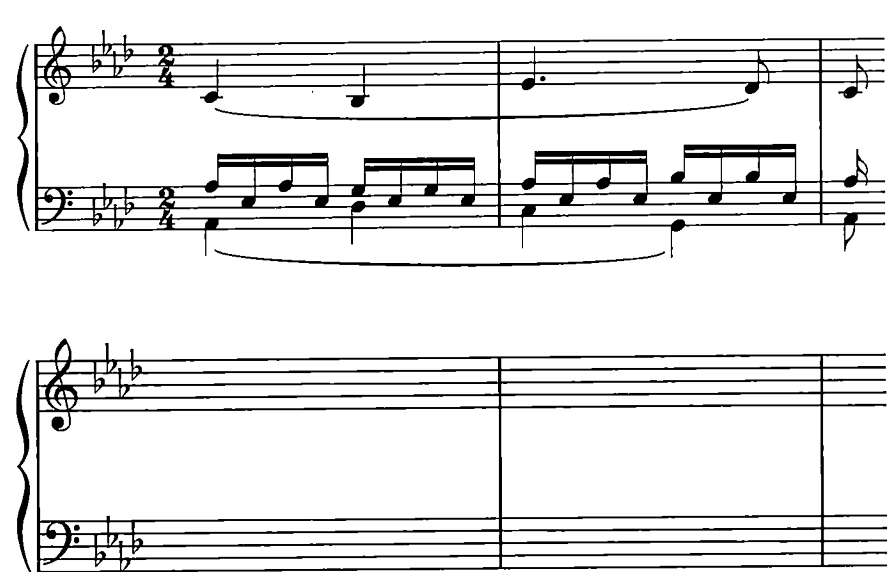

B. Which voice provides the greatest melodic interest? Of what kind of piece does this opening section remind you? To confirm your opinion, you might wish to check Beethoven's own title for this movement.

BEETHOVEN: PIANO SONATA IN A♭ MAJOR, OP. 26, III

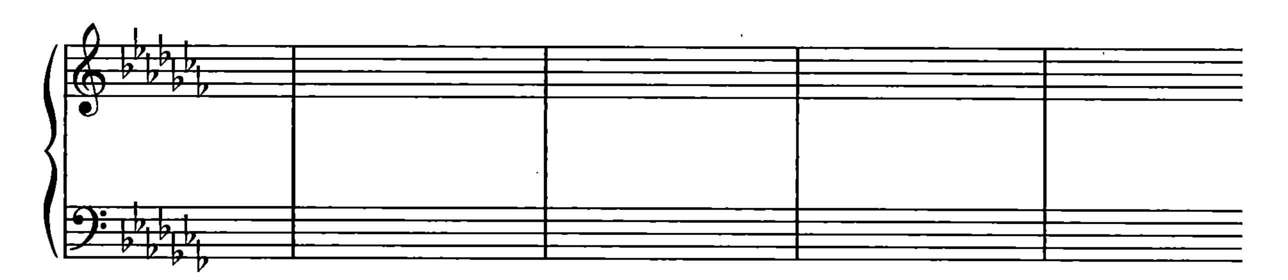

C. How is the chordal 7th of the V^7 in measure 2 approached?

BEETHOVEN: PIANO SONATA IN D MAJOR, OP. 2, NO. 2, II

6 Compose for piano an original parallel period (4 + 4 measures) that uses only tonic and dominant triads. You may select your own texture, but regardless of the texture that you choose, continue to employ the partwriting procedures discussed in the preceding chapters of the text. Keep your rate of harmonic change slow (about one harmony per measure). You may use the motive below or choose your own melodic ideas.

MOTIVE:

NAME ______________________________

CHAPTER 14

Melodic Figuration and Dissonance II

SUSPENSIONS AND SIMULTANEOUS DISSONANCES

1 Recopy the three-note suspension figure below, and add another part to produce 7–6, 4–3, and 2–3 suspensions in two-voice texture.

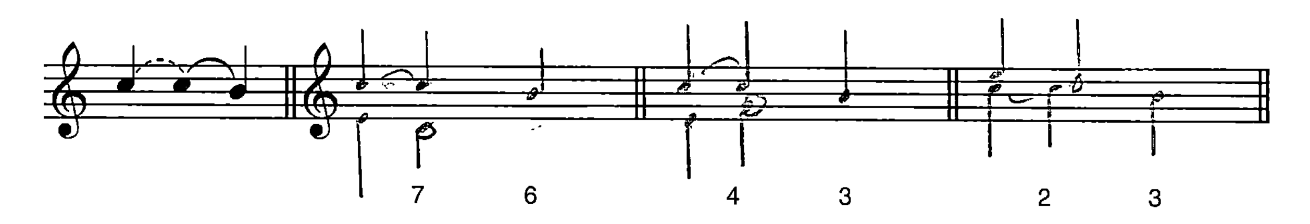

2 Recopy the two-voice passages below, adding appropriate suspensions in the upper or lower voice on those beats marked with an arrow. Denote each suspension with figured-bass symbols. Try to use some changes of bass (or upper part) in Exercise 2B.

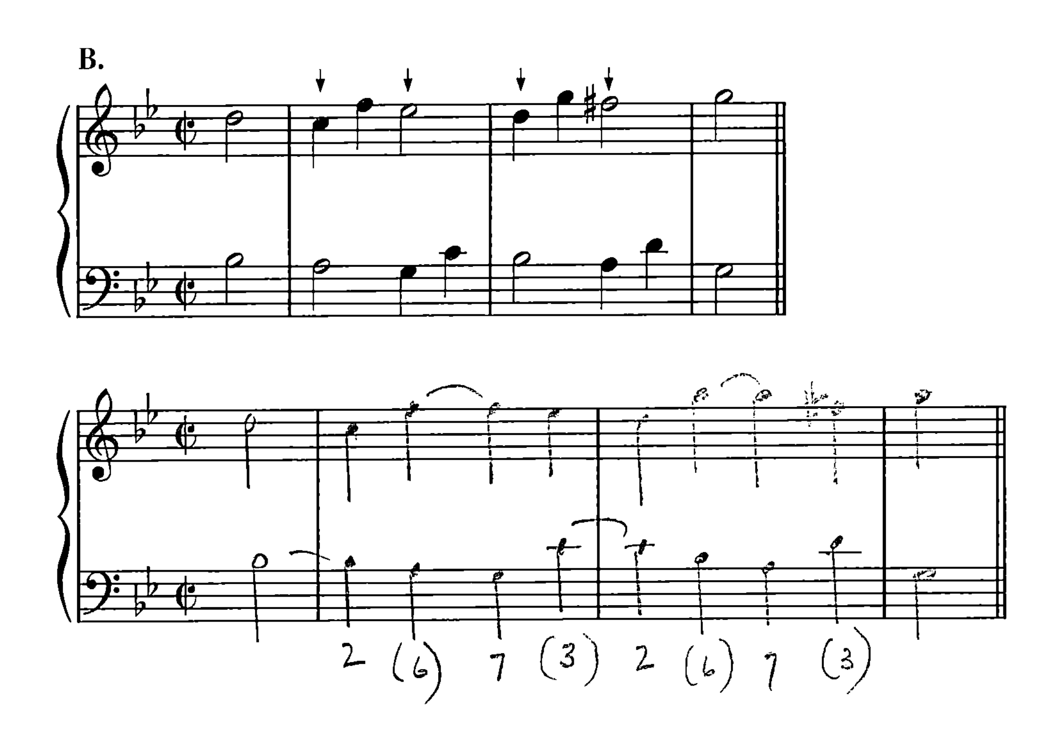

3 Realize the following figured-bass exercises in four-voice texture and provide a roman-numeral analysis for each below the staff. Be sure to distinguish between essential and embellishing harmonies. Remember that you do not double the resolution note in 7–6, 4–3, or 2–3 suspensions. Exercise 3B is unfigured, but more than one suspension is suggested by the outer voices. In Exercise 3C, the figures at the beginning of the second full measure indicate to double the 5th above the bass; possible rhythmic patterns for one or both inner voices are suggested above the staff. Be especially careful about what notes you double in Exercise 3D to avoid parallel perfect intervals.

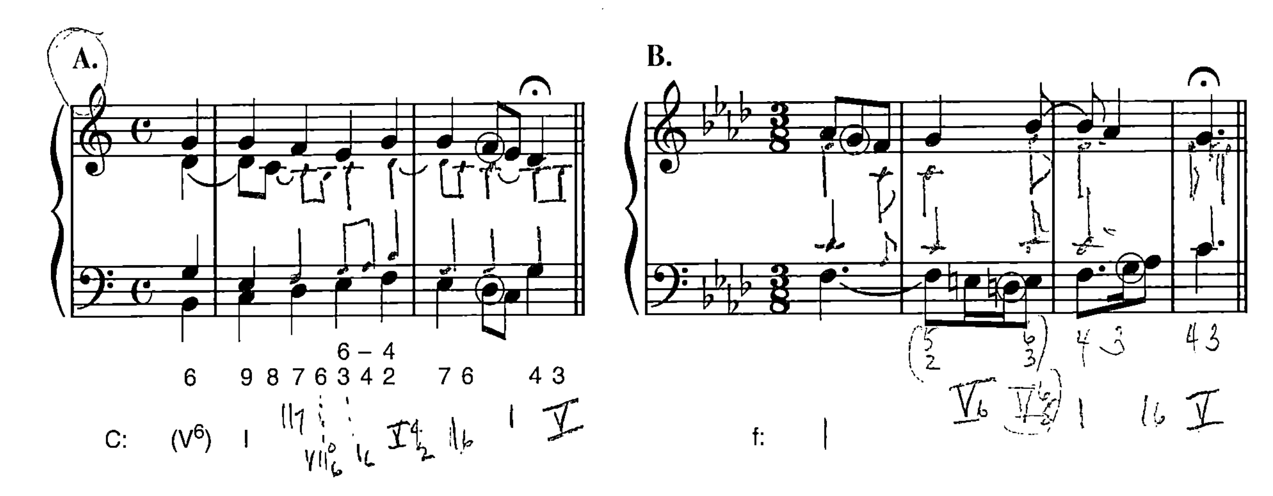

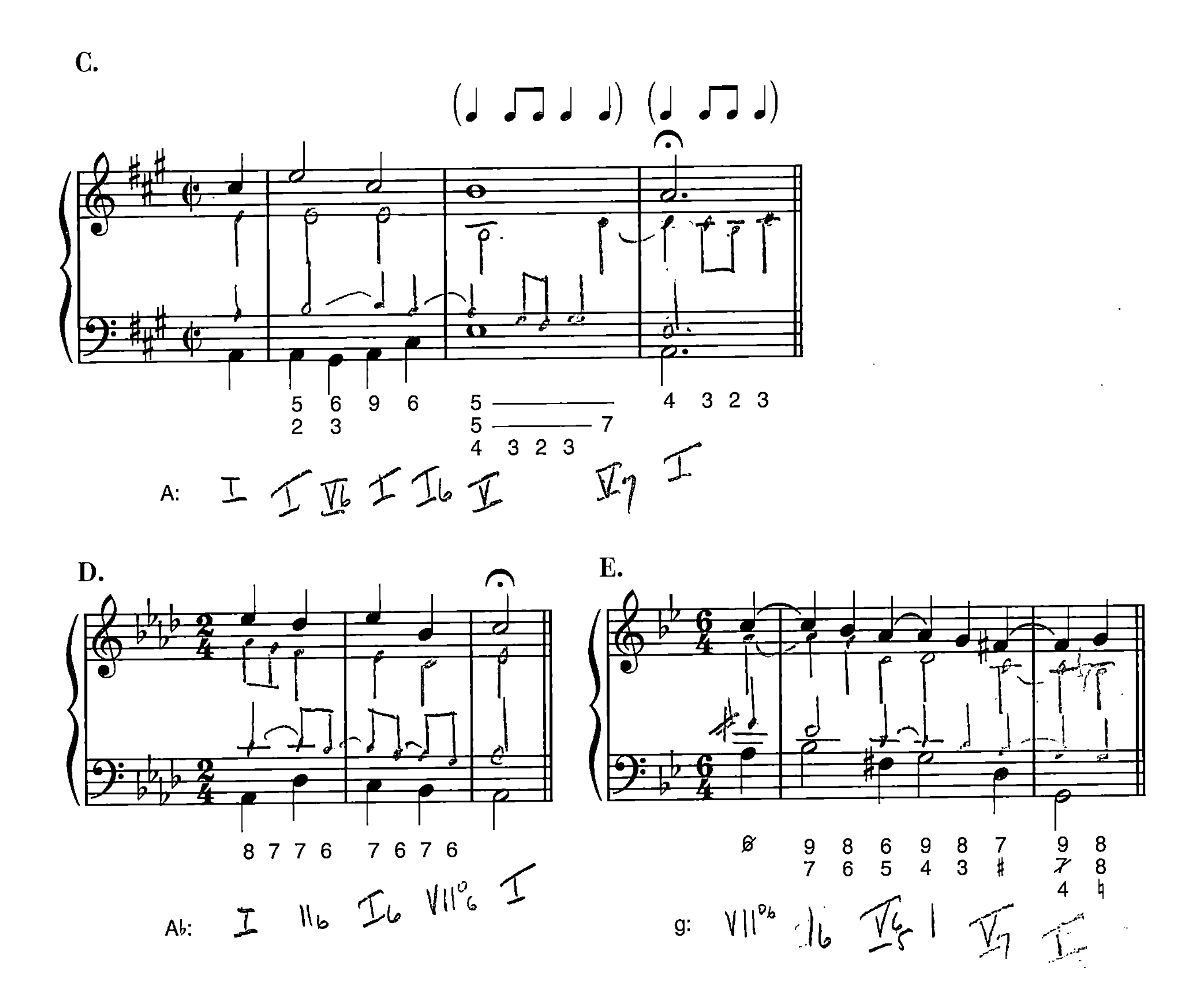

4 Given the following bass lines, figured-bass symbols, and rhythmic patterns above some staves, write out the indicated suspensions in four-voice texture; Exercise 4A is started for you. Then provide a roman-numeral analysis.

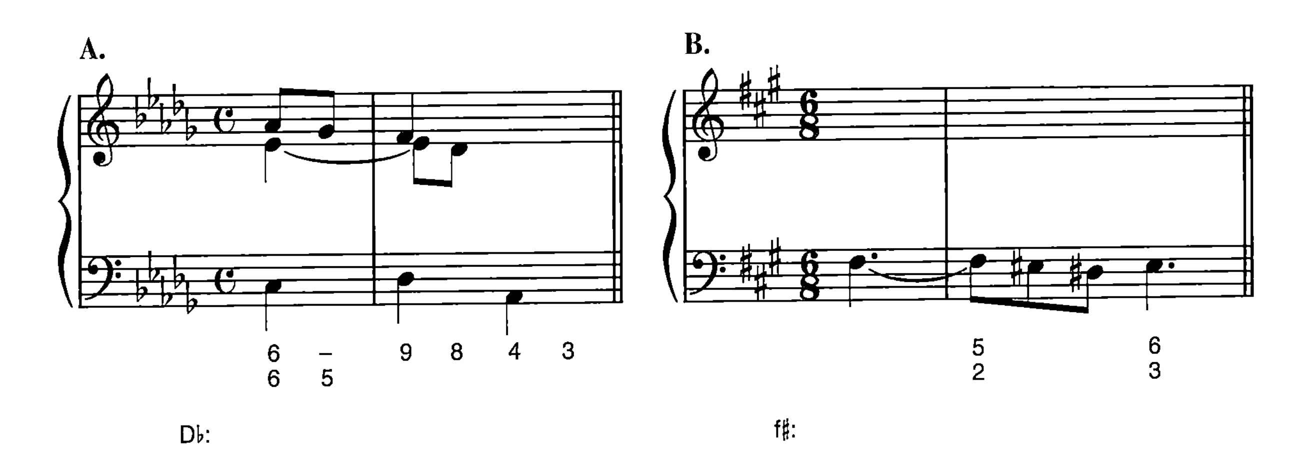

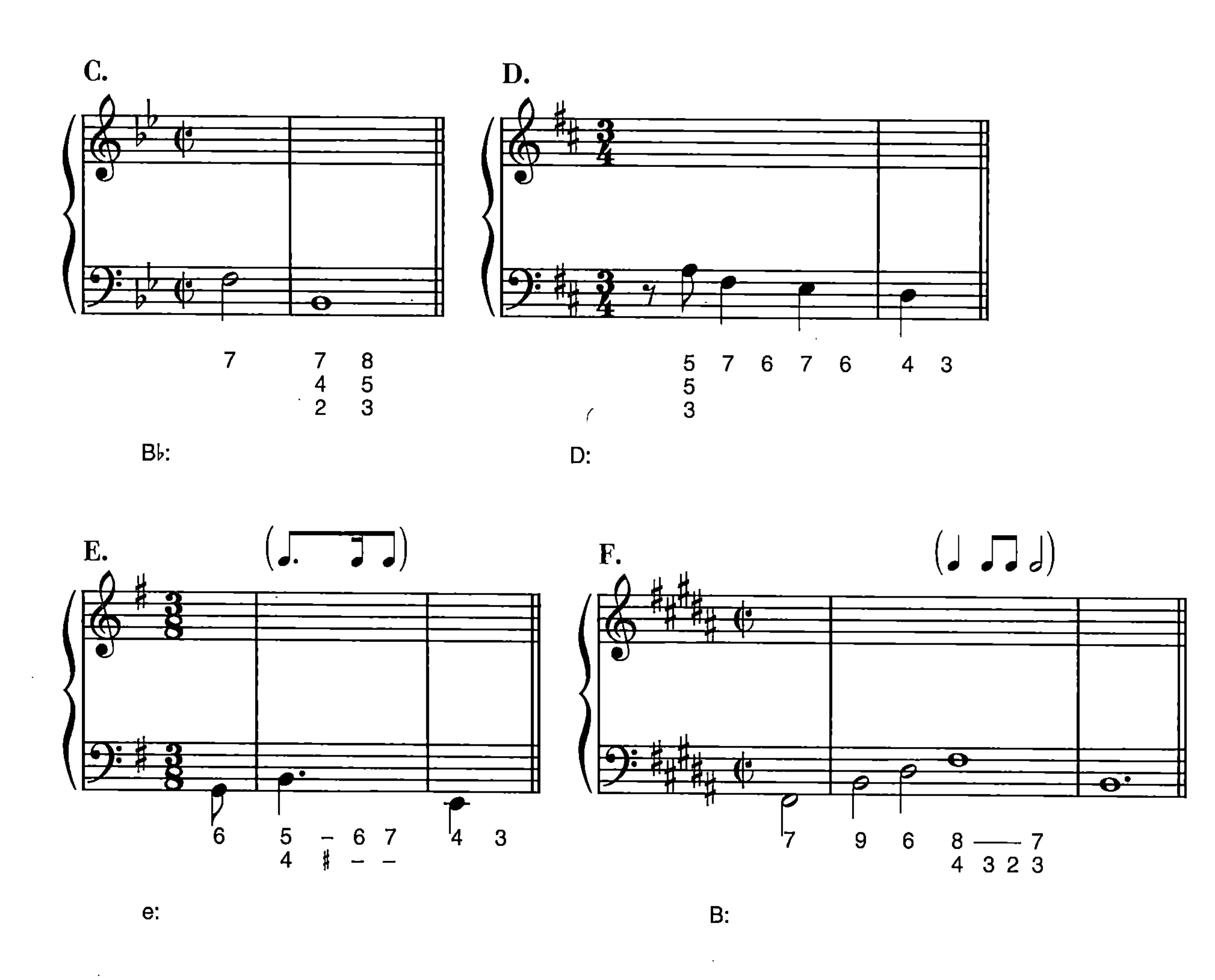

5 Harmonize the following two melodic lines in a four-voice setting, using *only* tonic and embellishing dominant-family harmonies. Add a good variety of appropriate suspensions. The chords change once per measure in the first tune; they move at a faster rate in the chorale setting.

B. "JESU, DEINE TIEFEN WUNDEN"

B♭:

6 Identify the various suspensions in the following excerpts by notating the correct symbols below the bass: 4–3, 7–6, etc. It is not necessary to analyze the harmonies.

A. What do you notice about the suspension resolutions?

MOZART: RONDO IN F MAJOR, K.494

B. In what two ways could the suspension in the last measure of this excerpt be analyzed?

THOMAS ARNE: THREE-VOICE CANON

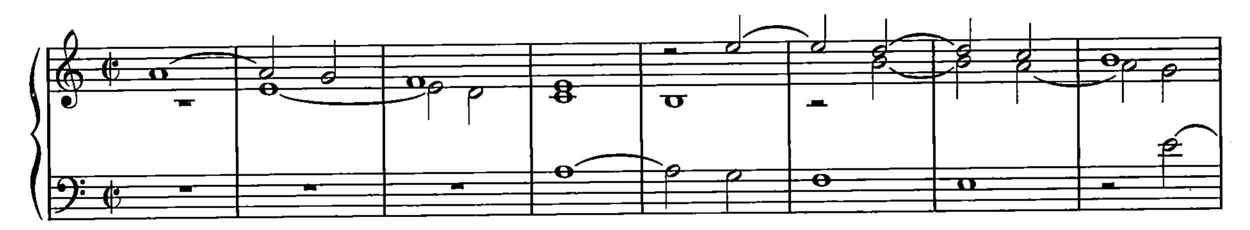

C. On the staff provided, reduce this compound melody to its constituent parts in order to reveal the suspensions within its implied two voices.

COUPERIN: "LA MILORDINE" FROM *PREMIÈRE ORDRE*

D. Make a simplification or reductive sketch of this passage before indicating the various suspensions. Indicate changes of bass or part.

BACH: PRELUDE IN B MINOR FROM *WELL-TEMPERED CLAVIER*, BOOK I

NAME ____________________

CHAPTER 15

Pre-Dominant Chords

IV AND II

1 Some cadential progressions are indicated below by roman numerals. Write out each bass line, and then add a soprano part; the first note of the soprano is given for each. Be sure to use appropriate scale degrees in the soprano at the cadence. Finally, fill in the inner voices.

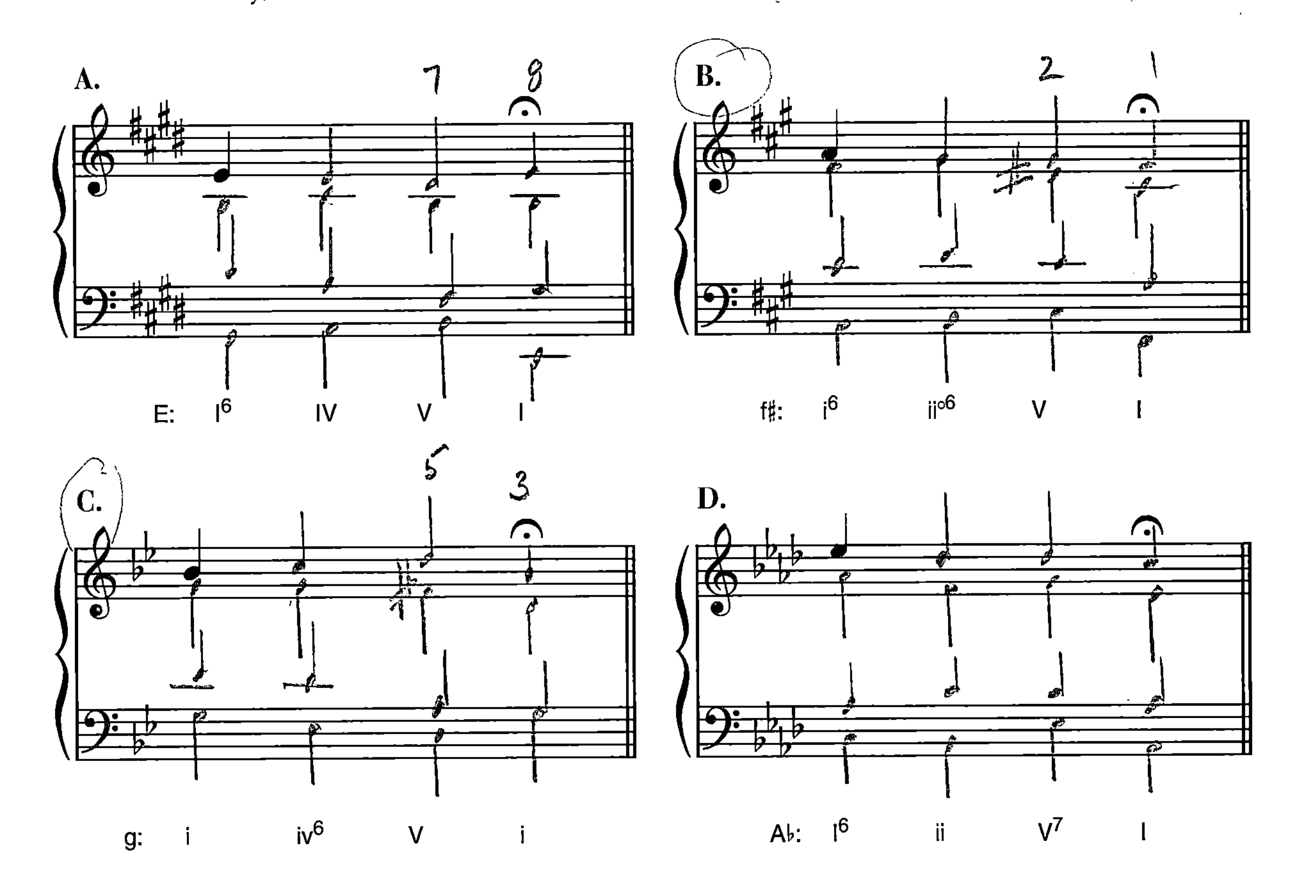

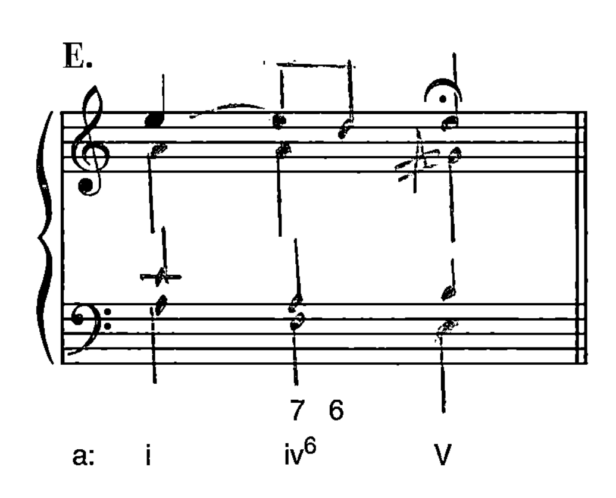

2 The following two melody harmonizations contain various partwriting errors. Circle the ones you find, and describe them briefly (citing measure numbers) in the spaces provided. What is peculiar about the ii° in the first measure of the second setting?

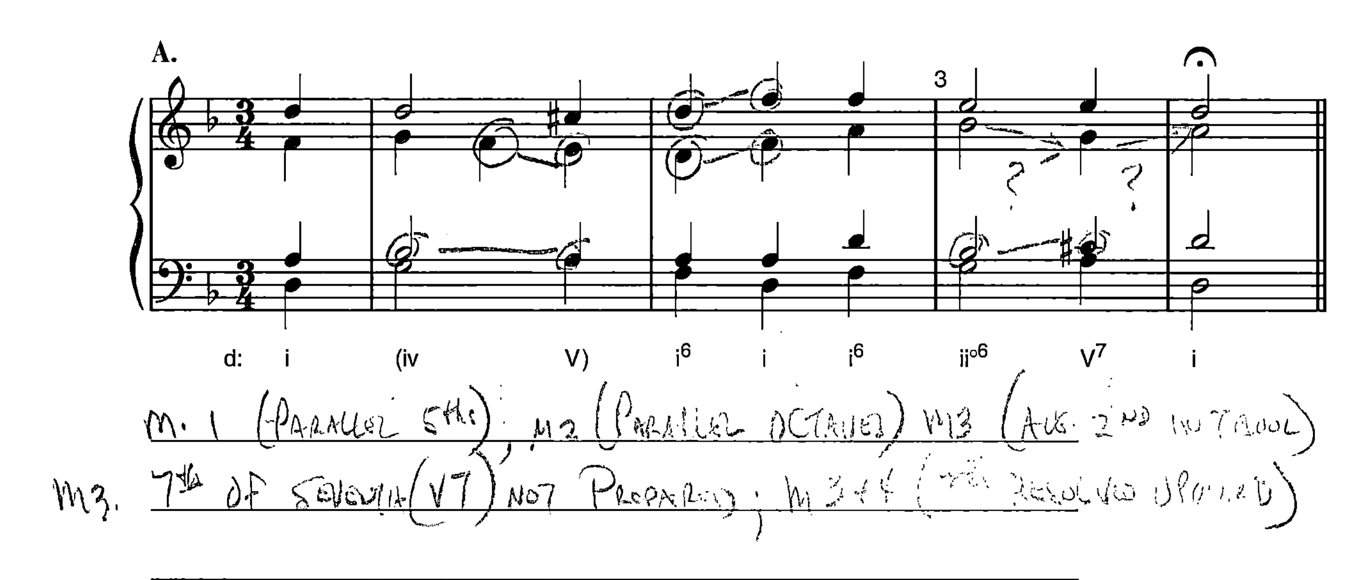

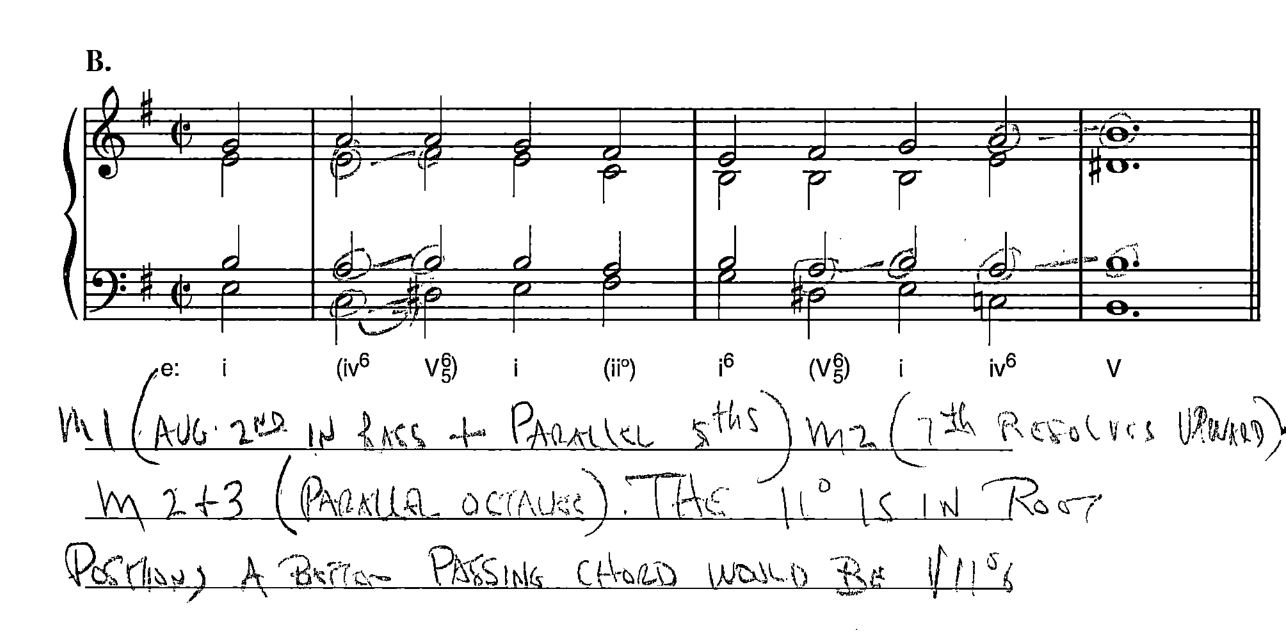

3 Realize the figured-bass exercises given below and provide a roman-numeral analysis below the staff. Key designations have not been provided here. Before you begin your roman-numeral analysis in these and the rest of the figured-bass exercises in this workbook, determine and indicate the key of each passage. Be careful to distinguish between essential and embellishing chords. Exercise 3D is unfigured.

A.

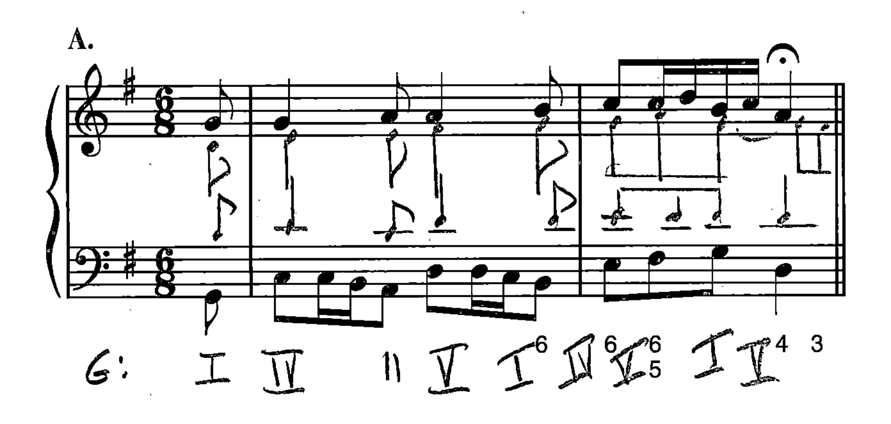

B.

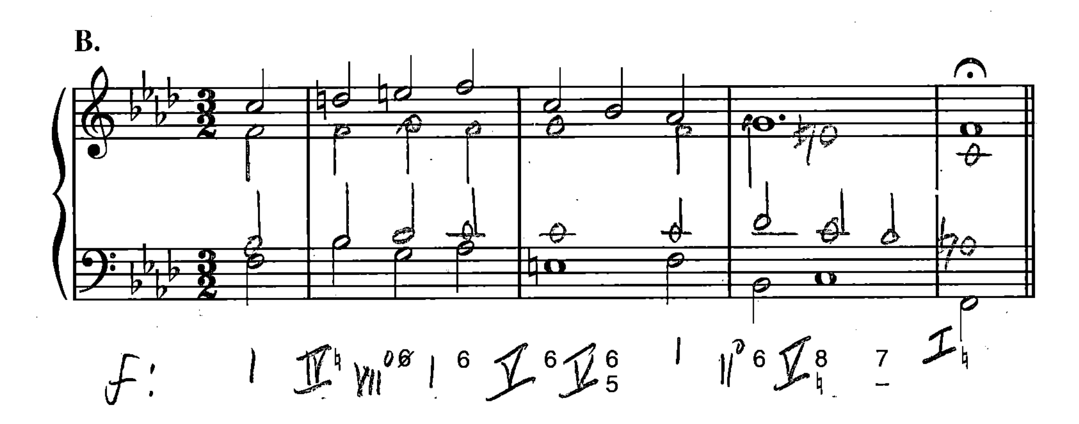

C.

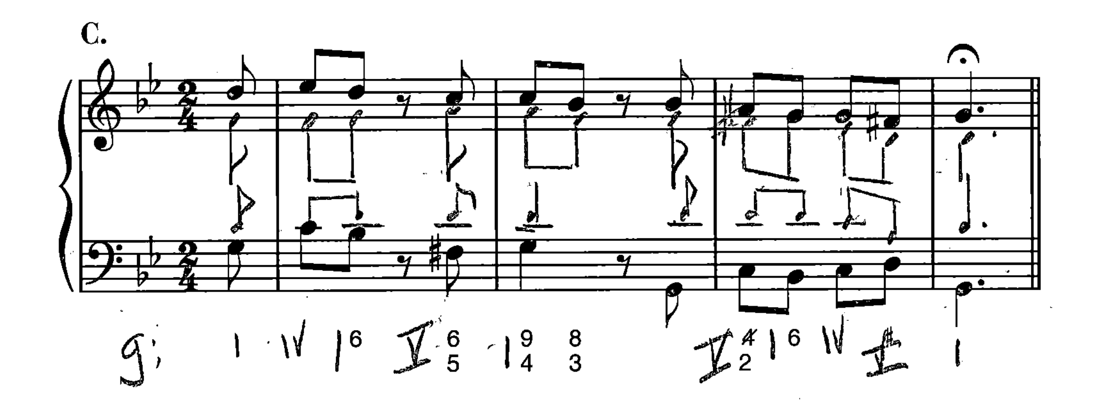

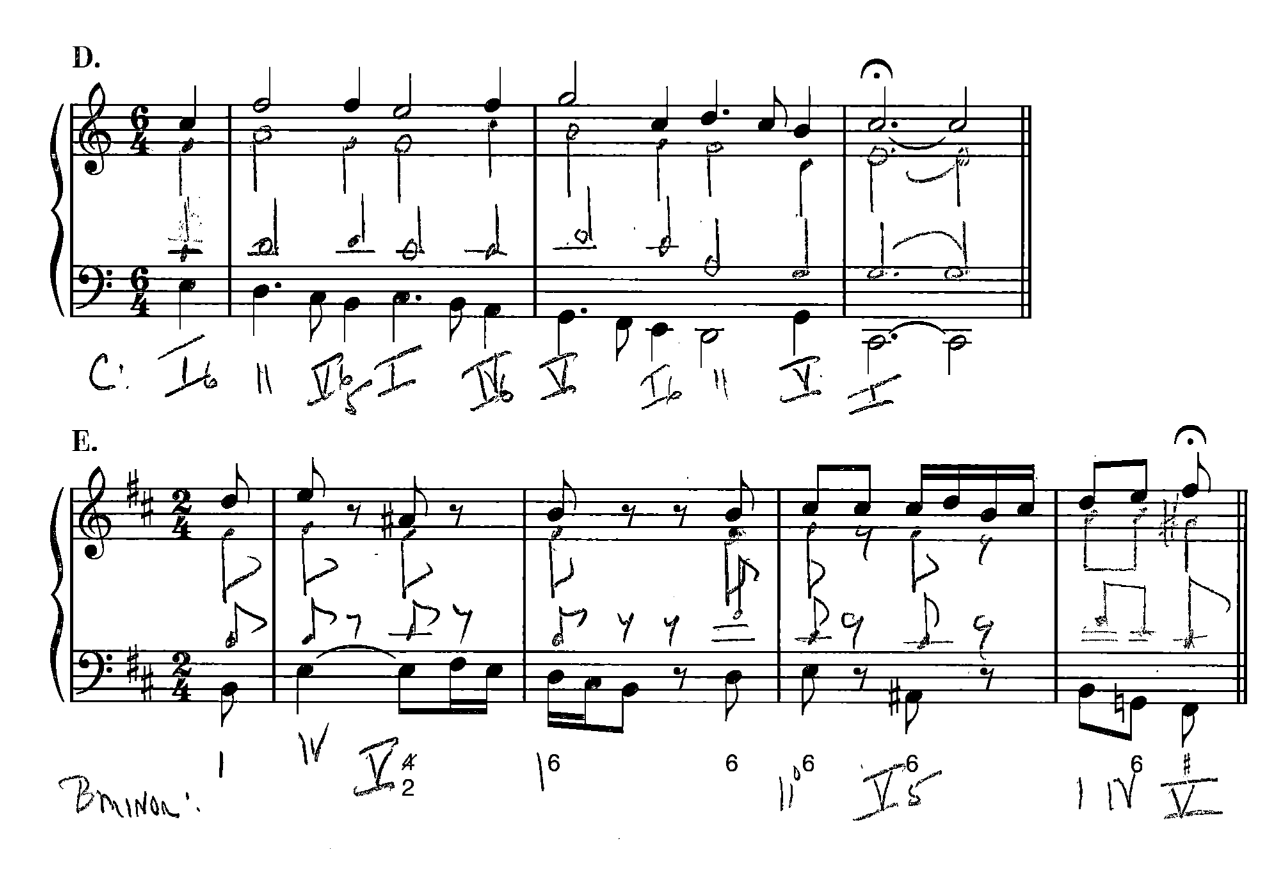

4 Make a four-voice setting of two of the following three melodies, employing a few appropriate pre-dominant chords. Give a roman-numeral analysis of your settings. Try to compose a line of running eighth notes for the bass in Exercise 4B; what type of cadence will you use? In Exercises 4A and 4B the chords should change almost every beat or quarter note; Exercise 4C demands a slower harmonic rhythm.

A. "O Haupt voll Blut und Wunden"

E♭:

B. "Wie schön leuchtet der Morgenstern"

C.

5 Make voice-leading reductions of the following phrases and give a roman-numeral analysis for each. In some cases, the entire phrase will suggest a cadential progression (such as I–IV–V–I), while in others you may find some embellishing chords within the phrase.

A. In your voice-leading reduction, begin your basic soprano voice on the G^5 in measure 1. To what note does the suspension in measure 2 resolve? What note in measure 3 continues this soprano line?

WEBER: CONCERTO FOR CLARINET IN E♭ MAJOR

B. What soprano note is supported by the bass G^2 in the third measure? Is this consistent with the previous measures?

VIVALDI: CONCERTO FOR FLUTE IN D MAJOR, II

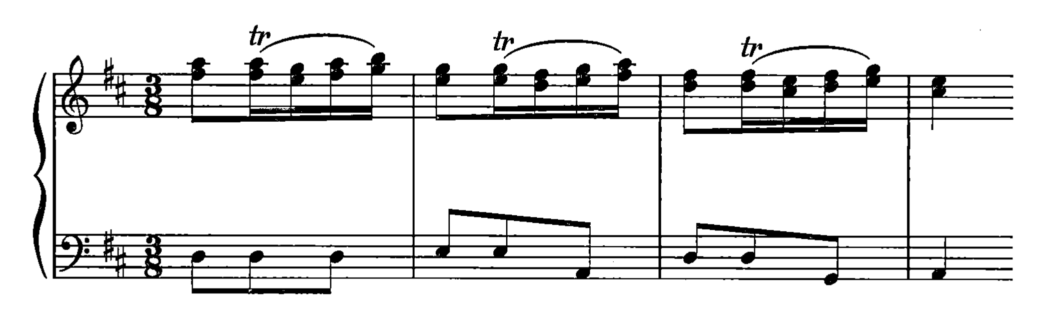

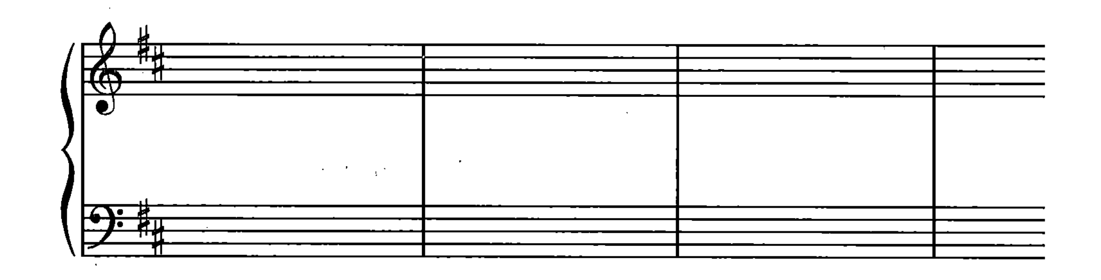

C. Consider this passage as one long phrase. Which subdominant chord is embellishing and which is cadential?

"ST. THEODULPH" (HYMN TUNE)

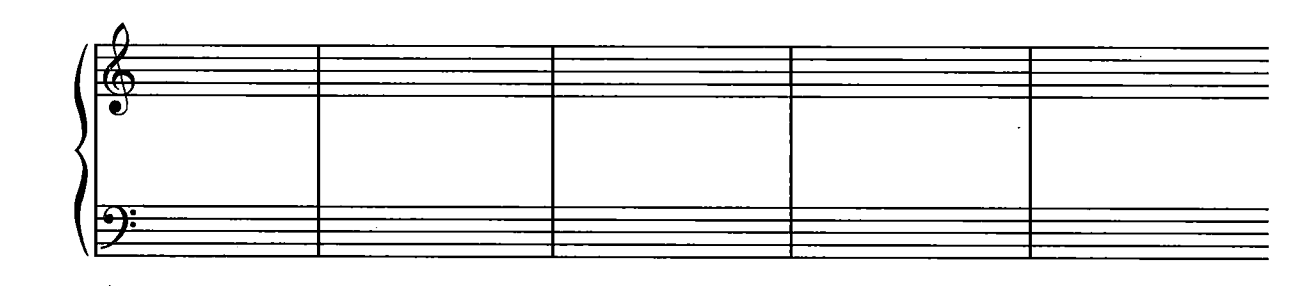

D. What happens on beat 2 of the opening measure? How do we know that (I) embellishes IV in measures 1–2? To which notes do the E^4 and E^3 move in measures 3 and 4? How do you account for the five-measure length of this phrase?

BRAHMS: SYMPHONY NO. 3 IN F MAJOR, II

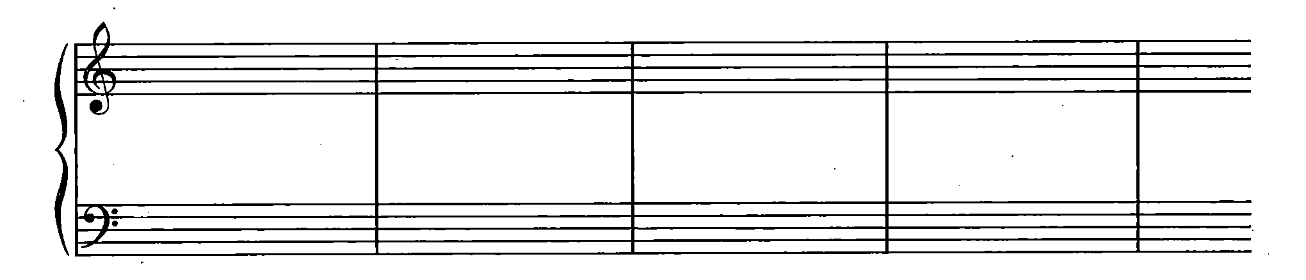

E. This excerpt begins with the last note of the preceding phrase. How does Bach elaborate the final dominant harmony?

"ES SPRICHT DER UNWEISEN MUND" (BACH CHORALE HARMONIZATION)

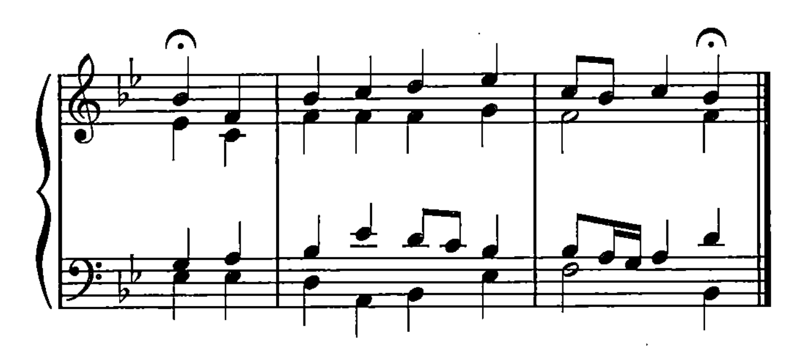

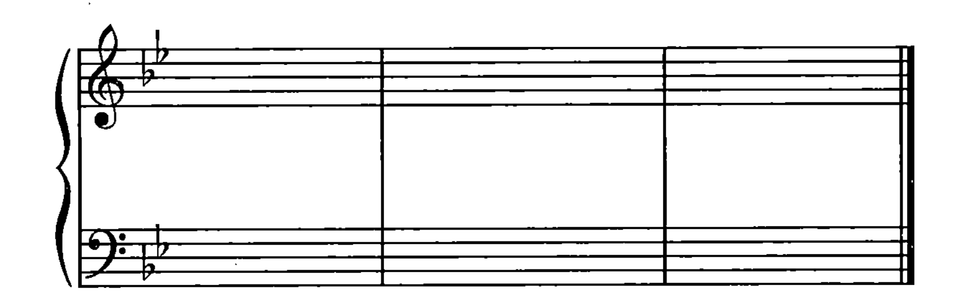

6 Supply a roman-numeral analysis for this short Chopin excerpt. Be cautious when indicating the use of the A♭ C E♭ triad, as it may function in different ways. Circle the notes of the ascending stepwise line in the soprano voice. The submediant triad (vi) that substitutes for the tonic in measure 4 prevents a premature cadence on the second beat.

CHOPIN: NOCTURNE IN G MINOR, OP. 37, NO. 1

NAME ____________________

CHAPTER 16

The $^6_4$ and Other Linear Chords

1 The hymn setting below contains various types of $^6_4$ chords. Circle each one, and identify its basic treatment by one of the following abbreviations: C (cadential), P (passing), N (neighboring), or A (arpeggiated). How would you explain the chord on the second eighth note of measure 1?

"HOLINESS" (AMERICAN HYMN TUNE)

2 Realize the following figured-bass exercises. Label each type of $^{6}_{4}$ chord using the abbreviations listed above. Supply a roman-numeral analysis, taking care to distinguish between essential and embellishing chords. Exercise 2B is unfigured.

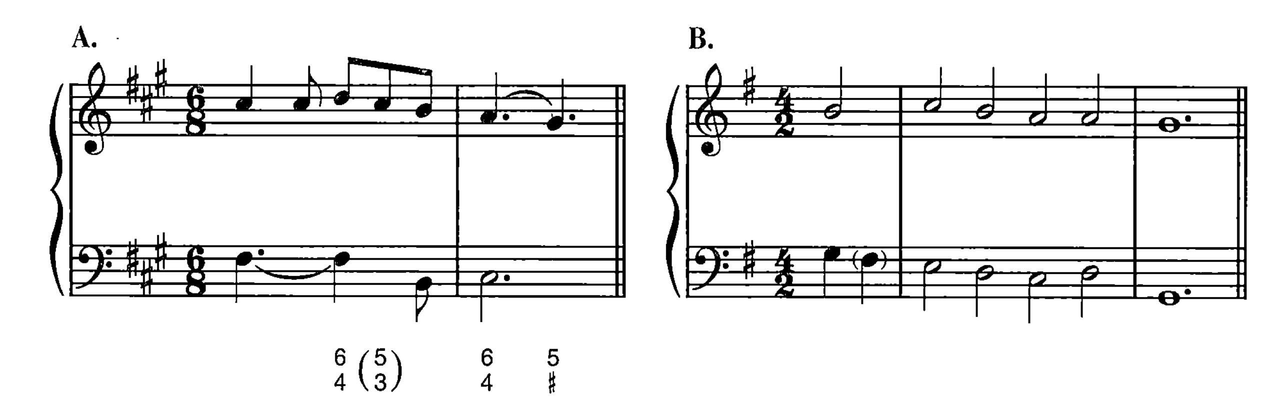

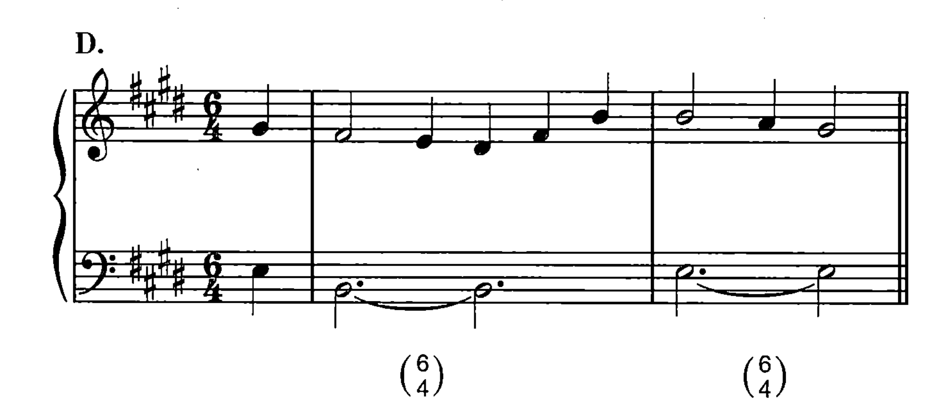

3 Harmonize the three short melodic fragments below, and insert a different type of $^6_4$ chord for each of the notes indicated with an arrow.

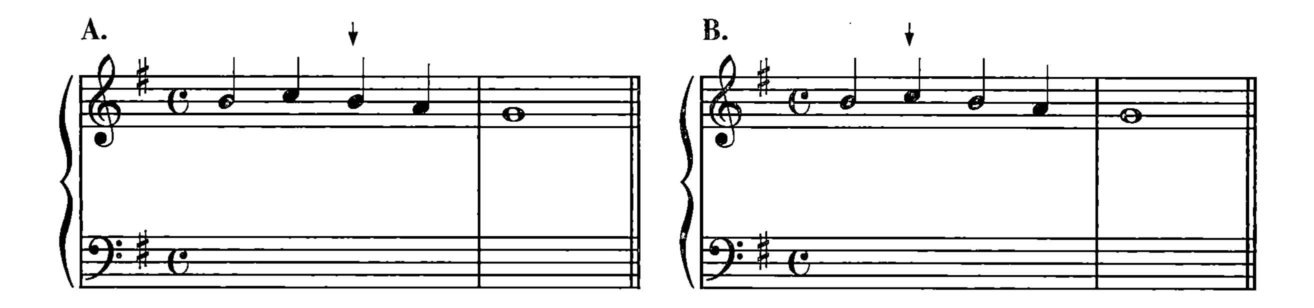

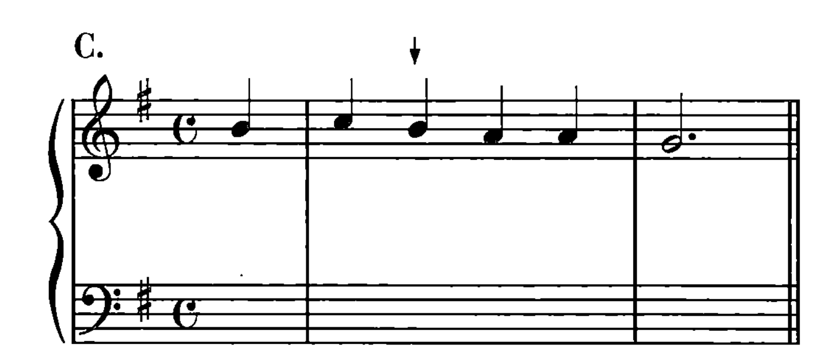

4 Make voice-leading reductions of the following excerpts, and then provide a roman-numeral analysis of each. Be sure to identify the various types of $^6_4$ chords that are employed.

A. What voice-leading device appears between the soprano and upper voice of the accompaniment in measures 1–2?

HAYDN: PIANO SONATA IN C MAJOR, HOB. XVI, NO. 35, I

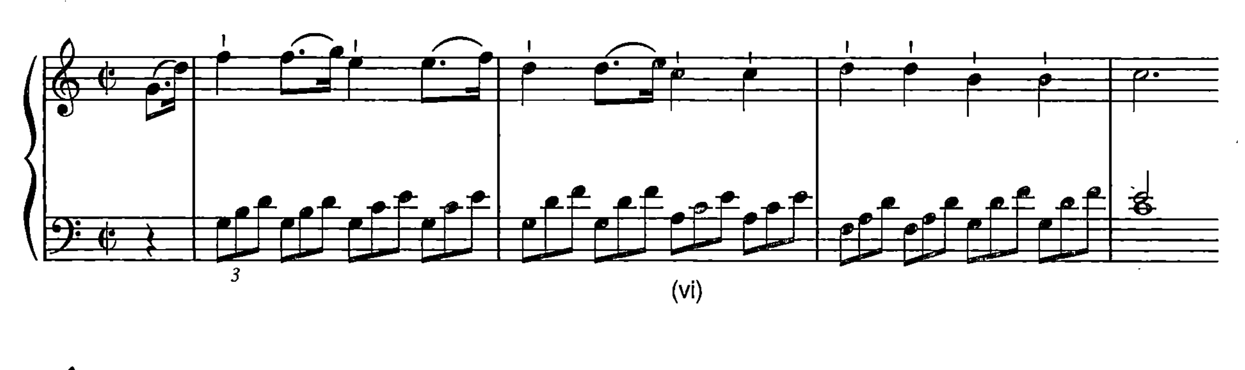

B. Is the same device used in the first two measures of this passage?

MOZART: PIANO SONATA IN C MAJOR K.330, III

C. What is similar about the use of the $\frac{6}{4}$ chord in the two excerpts below? What is different?

BEETHOVEN: PIANO SONATA IN F MINOR, OP. 2, NO. 1, II

NAME ______________________________

BRAHMS: VARIATIONS ON A THEME OF HANDEL, VAR. I

5 Provide a roman-numeral analysis, including $\substack{6\\4}$ chords, for each of the following two passages.

A. In what section of this Mendelssohn piece would the following passage most likely occur?

MENDELSSOHN: *SONGS WITHOUT WORDS*, OP. 38, NO. 1

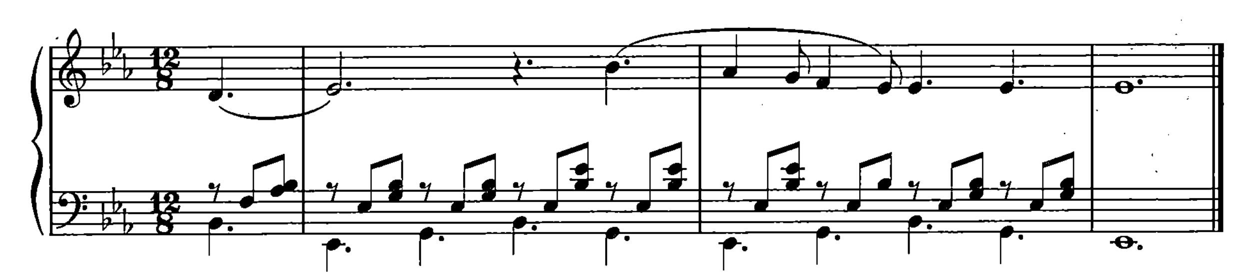

B. This opening passage from a rare Czech oratorio prominently features predominant harmonies. What is the function of the chords on the last two beats of measures 1 and 2 during this prolongation?

KRSTO ZYŽIK: *DIETA WORMSOVÁ*

6 Make a voice-leading reduction and roman-numeral analysis of the following passage; the first measure is done for you. Circle any instance of linear chords produced by passing motion, neighboring motion, etc. in the original passage.

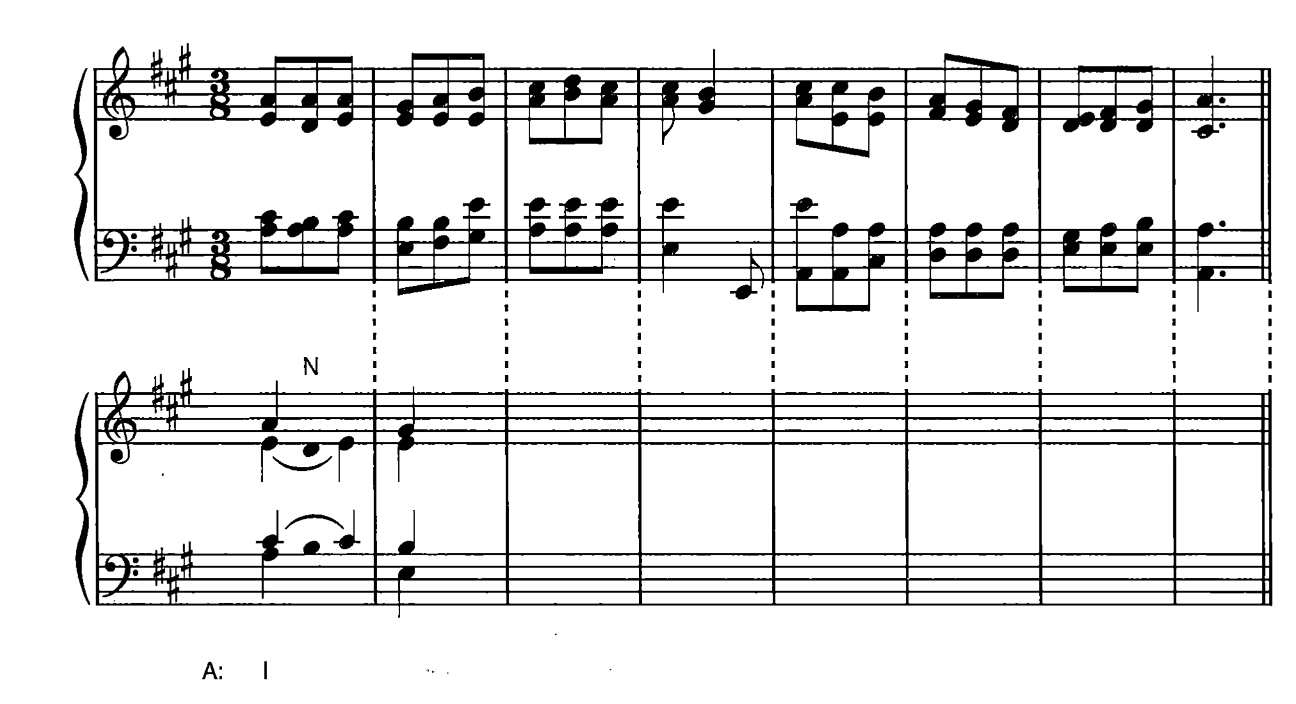

NAME ______________________________

CHAPTER 17

The II^7 and IV^7 Chords

1 Roman-numerals for several short progressions are provided below. Complete each progression in four-voice texture. Write out your bass line first, and then add an appropriate soprano line; the final soprano note is given in each case. Be sure that you approach and resolve the chordal 7th of each pre-dominant seventh chord properly. Can you describe how the progression in Exercise 1E is different from the others?

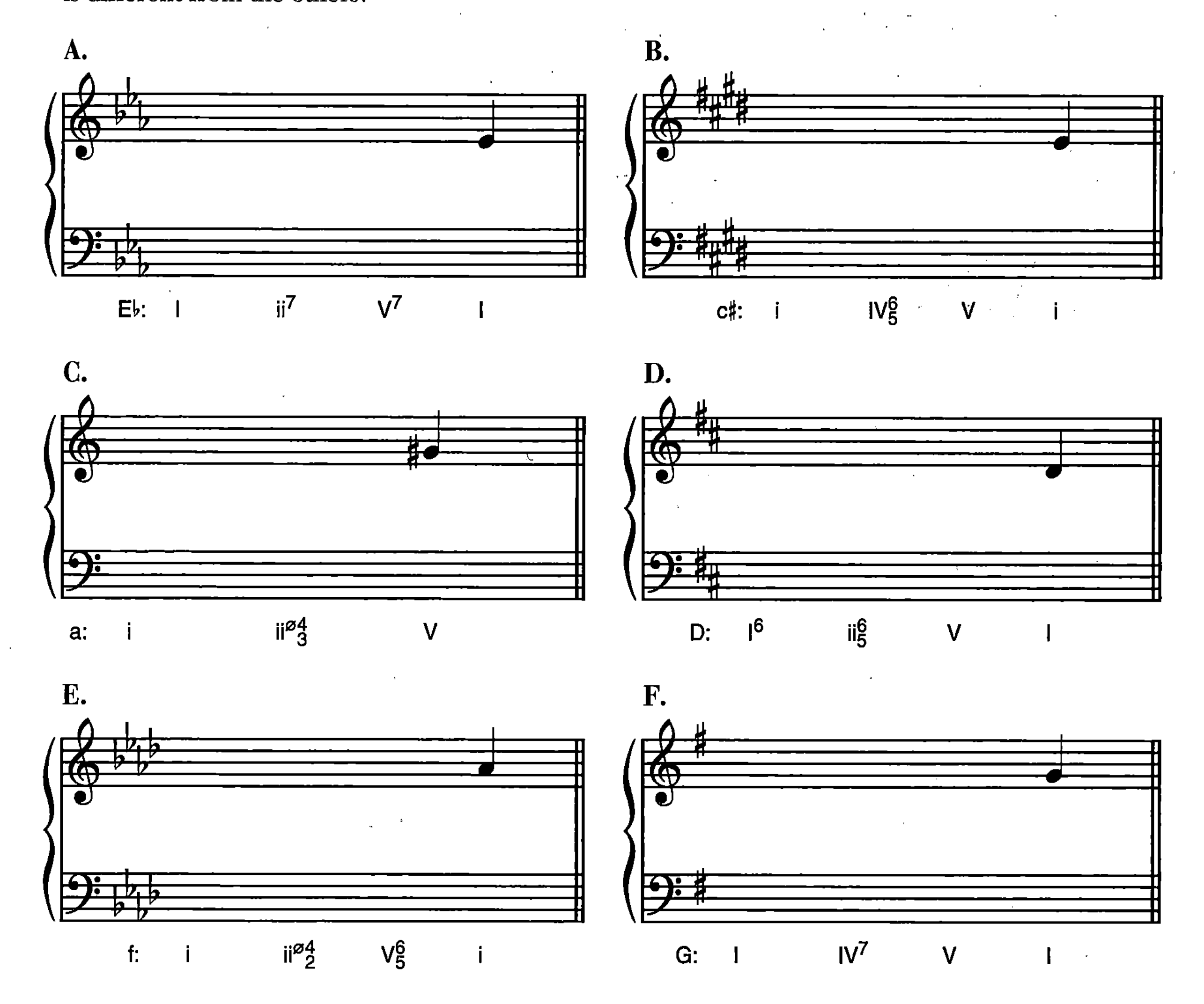

2 Supply roman-numeral analyses for the passages below, putting parentheses around all embellishing pre-dominant chords. For each pre-dominant seventh chord, indicate whether the chordal 7th is treated as a suspension (S), passing tone (P), or incomplete neighbor (IN). There is an example of parallel 5ths hidden in one of the passages; can you find it?

3 Realize the following figured-bass exercises, and supply roman-numeral analyses below the bass-clef staff. Explain the handling of the dissonant 7th in the pre-dominant seventh chords. Exercise 3D is unfigured; supply three appropriate seventh chords for it.

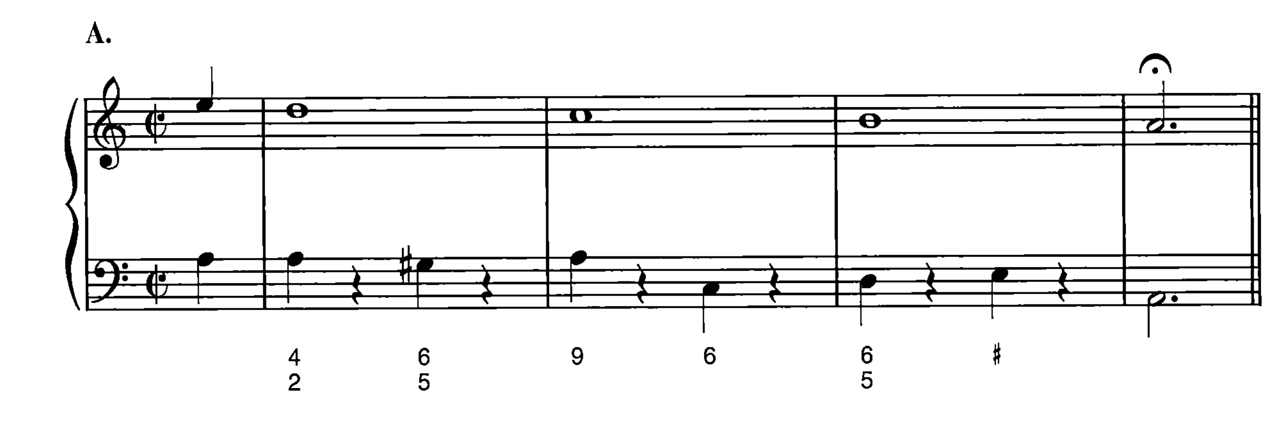

B. C.

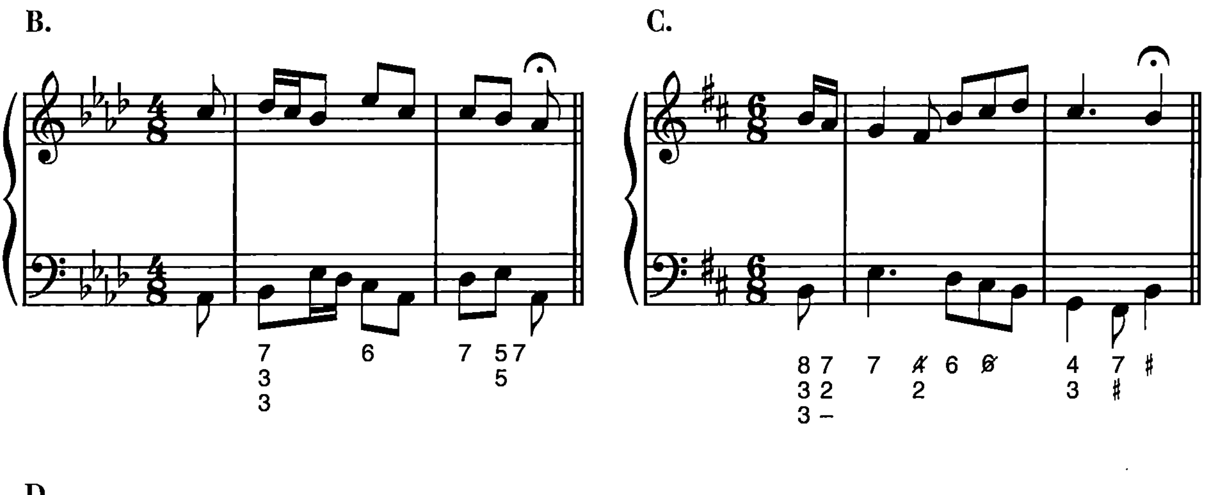

D.

4 Harmonize the melodic period below, using a different inversion of a predominant seventh chord on each note marked with an arrow. Strive for an interesting bass line. Then make a voice-leading reduction, and give a roman-numeral analysis of your finished piece.

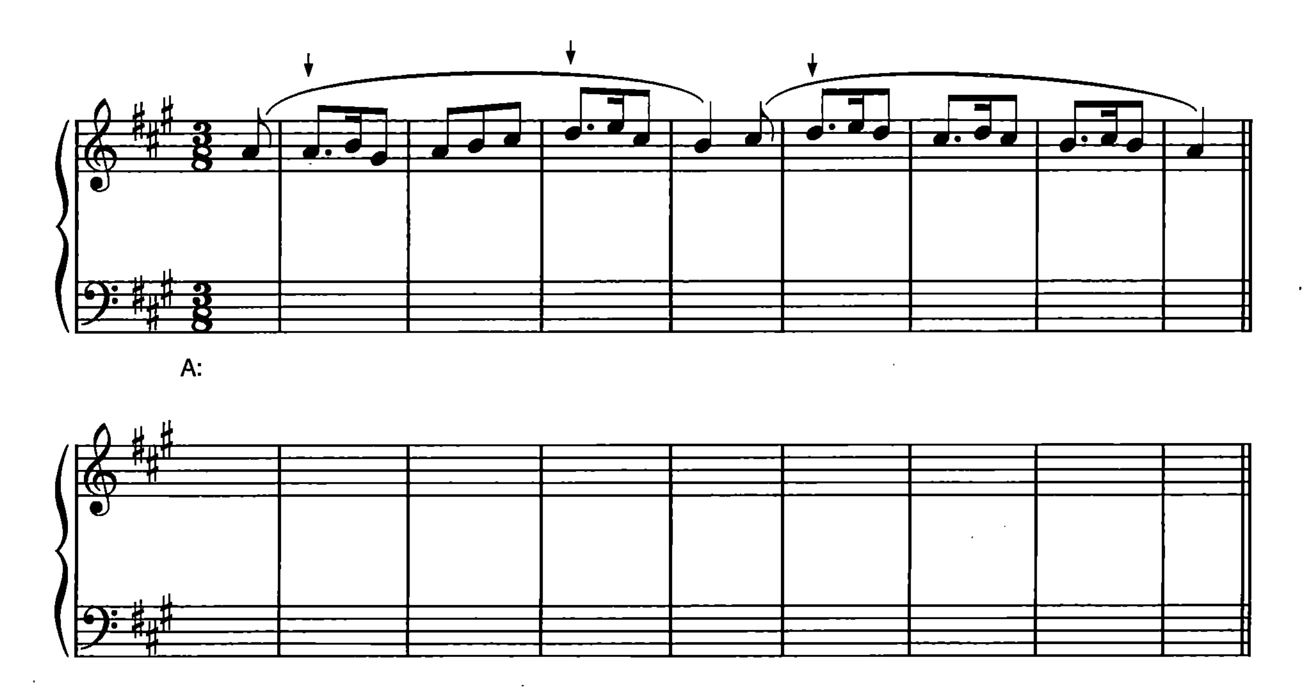

5 Analyze the following excerpts from music literature. Make voice-leading reductions of all passages except the Bach Sonata (Exercise 5F).

A. Only one complete measure of space is provided for your reduction. Why? To what chord does the first pre-dominant seventh chord "resolve"?

SCHUMANN: "****" FROM *ALBUM FOR THE YOUNG*, OP. 68, NO. 30

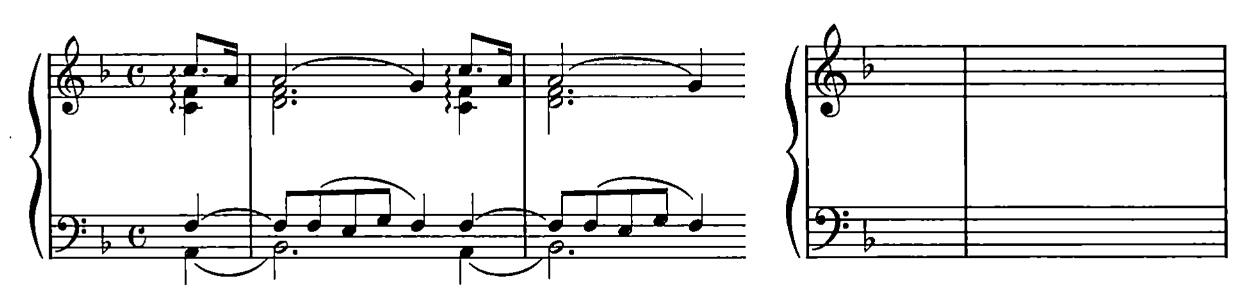

B. Why must the pre-dominant seventh chord be incomplete?

HANDEL: "LASCIA CH'IO PIANGA" FROM *RINALDO*, ACT II

NAME ______________________________

C. In which measure does a pre-dominant seventh chord appear in this famous horn tune? What is curious about it?

TCHAIKOVSKY: SYMPHONY NO. 5 IN E MINOR, II

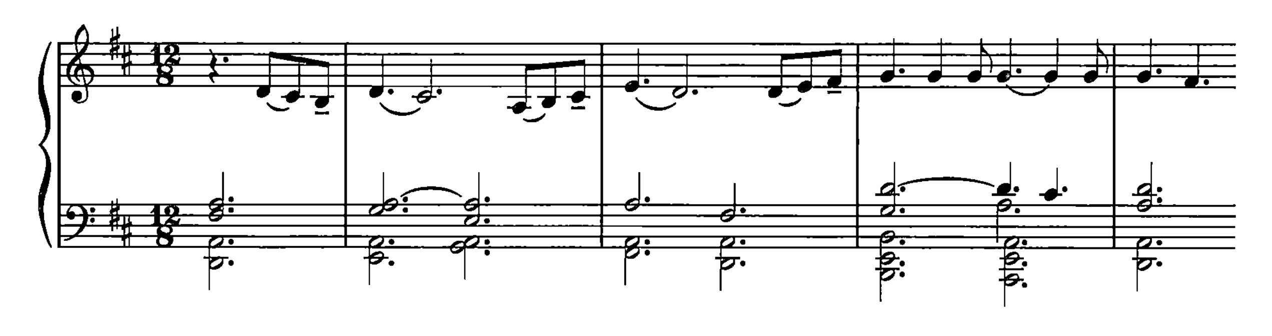

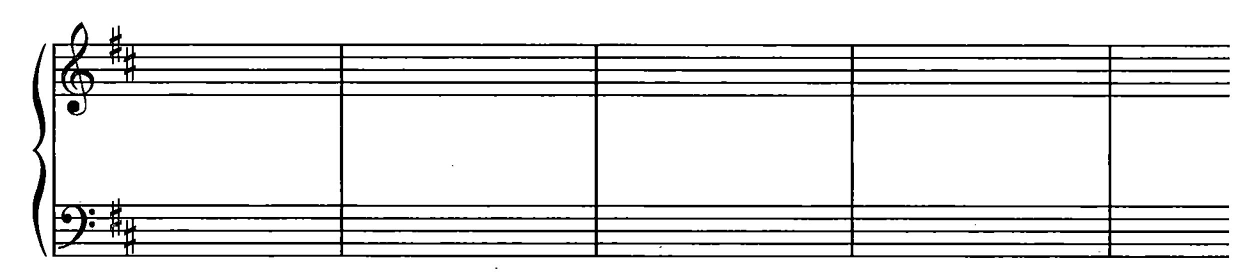

D. How does Corelli avoid parallels in measures 4–5?

CORELLI: CORRENTE FROM CONCERTO GROSSO NO. 10 IN C MAJOR

E. To what chord does the pre-dominant harmony resolve? What is unusual about the resolution of its 7th?

MOZART: PIANO SONATA IN F MAJOR, K.280, II

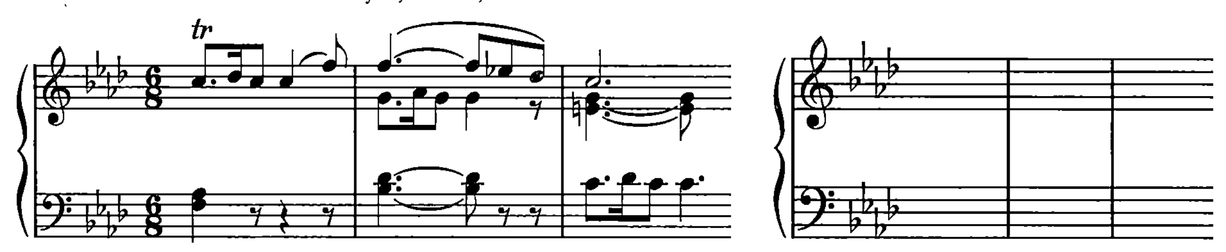

F. Does this passage represent a cadential or embellishing progression?

BACH: SONATA FOR FLUTE AND CLAVIER IN E♭ MAJOR, II

G. What is the function of the chords on the second quarter notes in measures 5–8?

BEETHOVEN: SYMPHONY NO. 2 IN D MAJOR, III

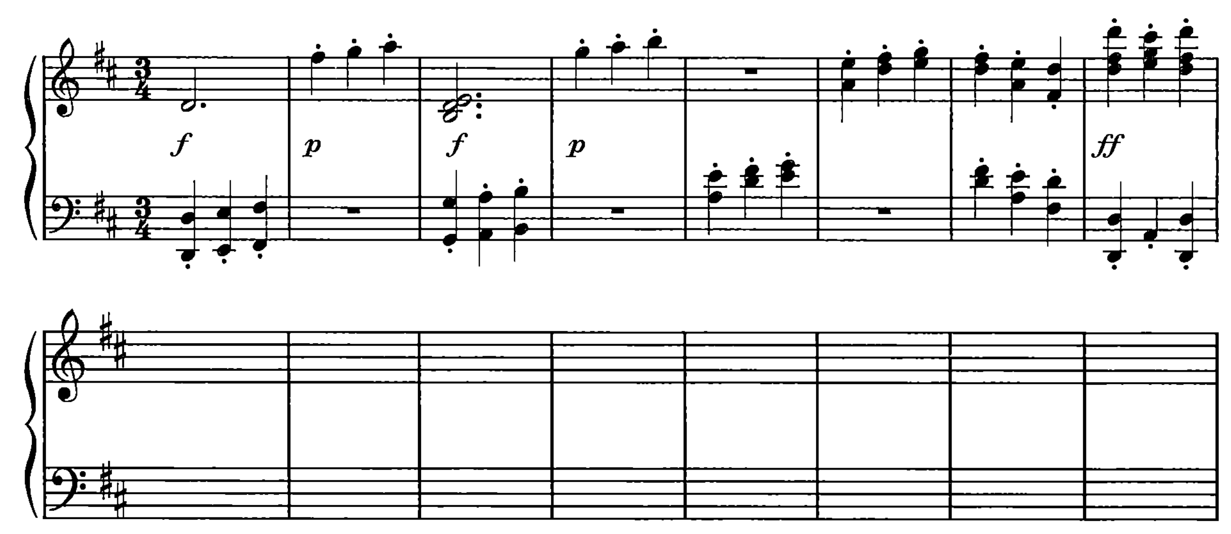

NAME ___________________________

CHAPTER 18

The VI, III, and Other Diatonic Chords

1 The soprano lines and roman-numerals are provided for the following progressions. First supply the correct bass lines, and then fill in the alto and tenor voices.

2 Realize the following figured-bass exercises, and provide a roman-numeral analysis below each bass-clef staff. Be especially careful not to create parallel perfect intervals in the opening measures of Exercises 2C and 2E. Identify any instances of mediant or submediant triads that substitute for either I or I^6.

E.

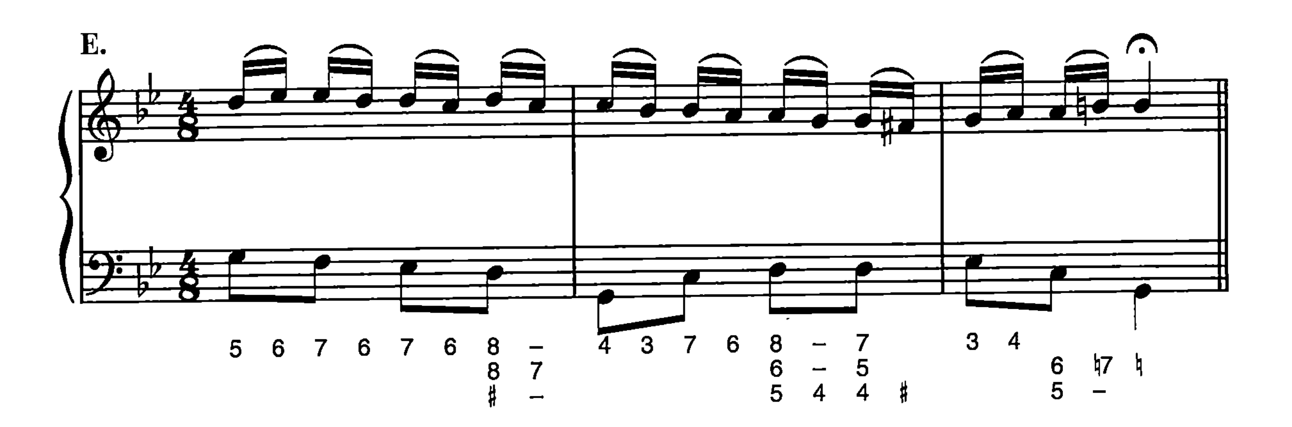

3 Harmonize the following two melodies. The hymn (Exercise 3A) suggests a chorale-like texture in four voices. You might prefer a freer texture in the folk song (Exercise 3B). Use at least one example of both a vi and a iii in each setting.

A. "OLD HUNDRED"

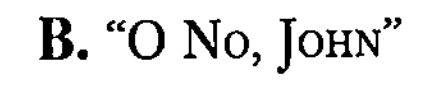

B. "O NO, JOHN"

4 Compose an original parallel period that employs the diatonic triads and seventh chords we have studied thus far. You may write either a three- or four-voice setting and occasionally use some octave doublings. Strive for stepwise melodic lines in the outer voices, and add a few nonharmonic tones where appropriate.

NAME ______________________________

5 Analyze the excerpts below, providing a voice-leading reduction and roman-numeral analysis for each one.

A. These two phrases open the first and last movements of Mozart's Clarinet Quintet. Both make prominent use of the submediant chord. After completing your analyses, compare the two harmonic schemes and be able to comment on any similarities between them.

MOZART: CLARINET QUINTET IN A MAJOR, K.581, I

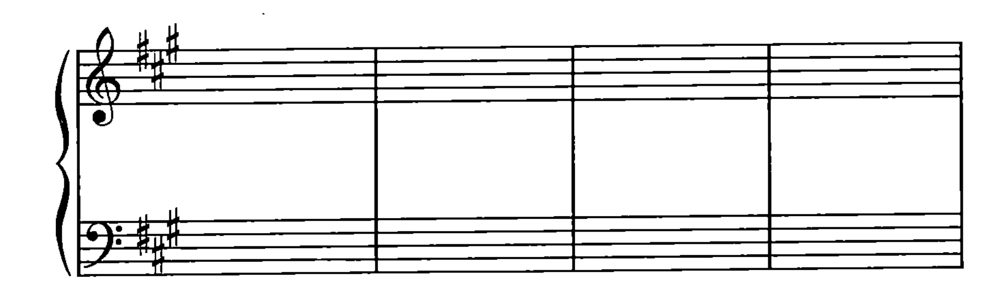

MOZART: CLARINET QUINTET IN A MAJOR, K.581, IV

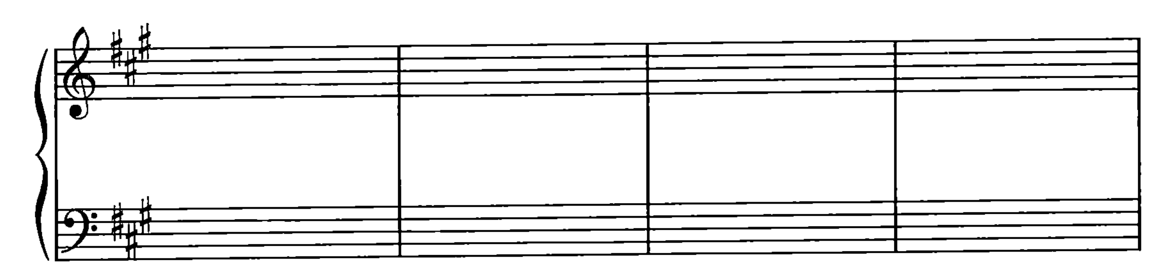

B. What is the function of the VI chord in this passage?

BACH: CHACONNE FROM VIOLIN PARTITA NO. 2 IN D MINOR

C. Before making your reduction, determine which chords are the essential harmonies of this passage.

TCHAIKOVSKY: SYMPHONY NO. 4 IN F MINOR, III

NAME ______________________________

D. The voice leading of this aria is particularly elegant. What happens in the soprano line of your reduction?

HANDEL: "VERDI PRATI" FROM *ALCINA*, ACT II

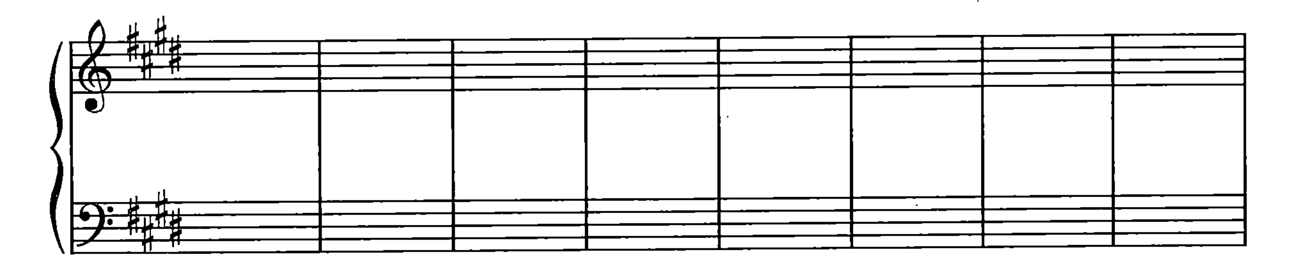

6 Supply a roman-numeral analysis for this excerpt. What is unusual about this progression? Why do you think Dvořák chose the particular harmony in measure 4?

DVOŘÁK: SYMPHONY NO. 9 IN E MINOR ("NEW WORLD"), I

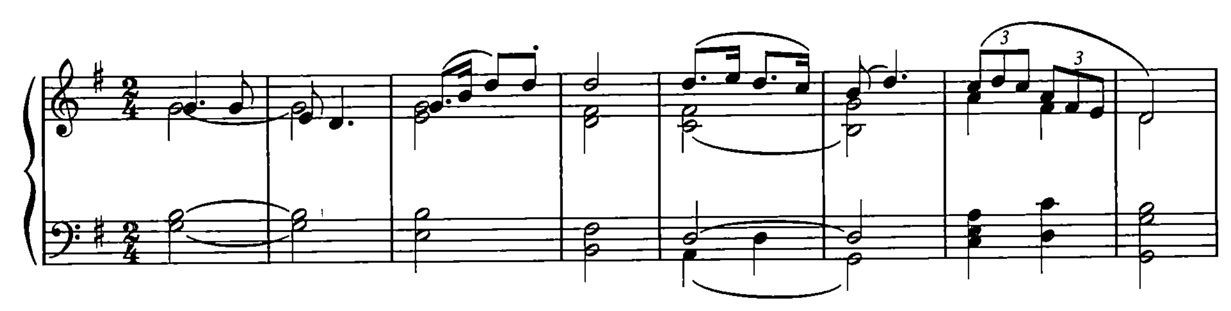

7 This chorale phrase opens in A minor and closes in a tonicized C major, as indicated by the brackets below the staff. Keeping these harmonies in mind, what type of seventh chord is found in measures 1 and 3 and how is each derived?

"PUER NATUS IN BETHLEHEM" (BACH CHORALE HARMONIZATION)

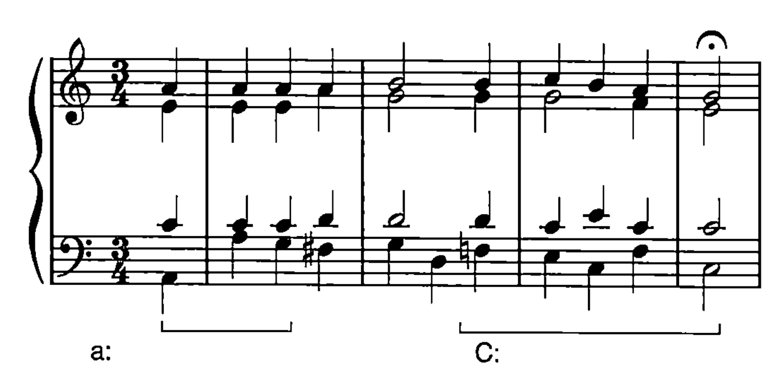

____

____

NAME ____________________

CHAPTER 19

Tonicization and Modulation I

MOTION TO V AND III

1 The following phrases modulate to the dominant [V] or relative major [III]. The original key, pivot chord, and outer voices of the final chord are supplied. Create progressions in four voices, and provide a roman-numeral analysis for each. What kind of modulation occurs in Exercise 1E?

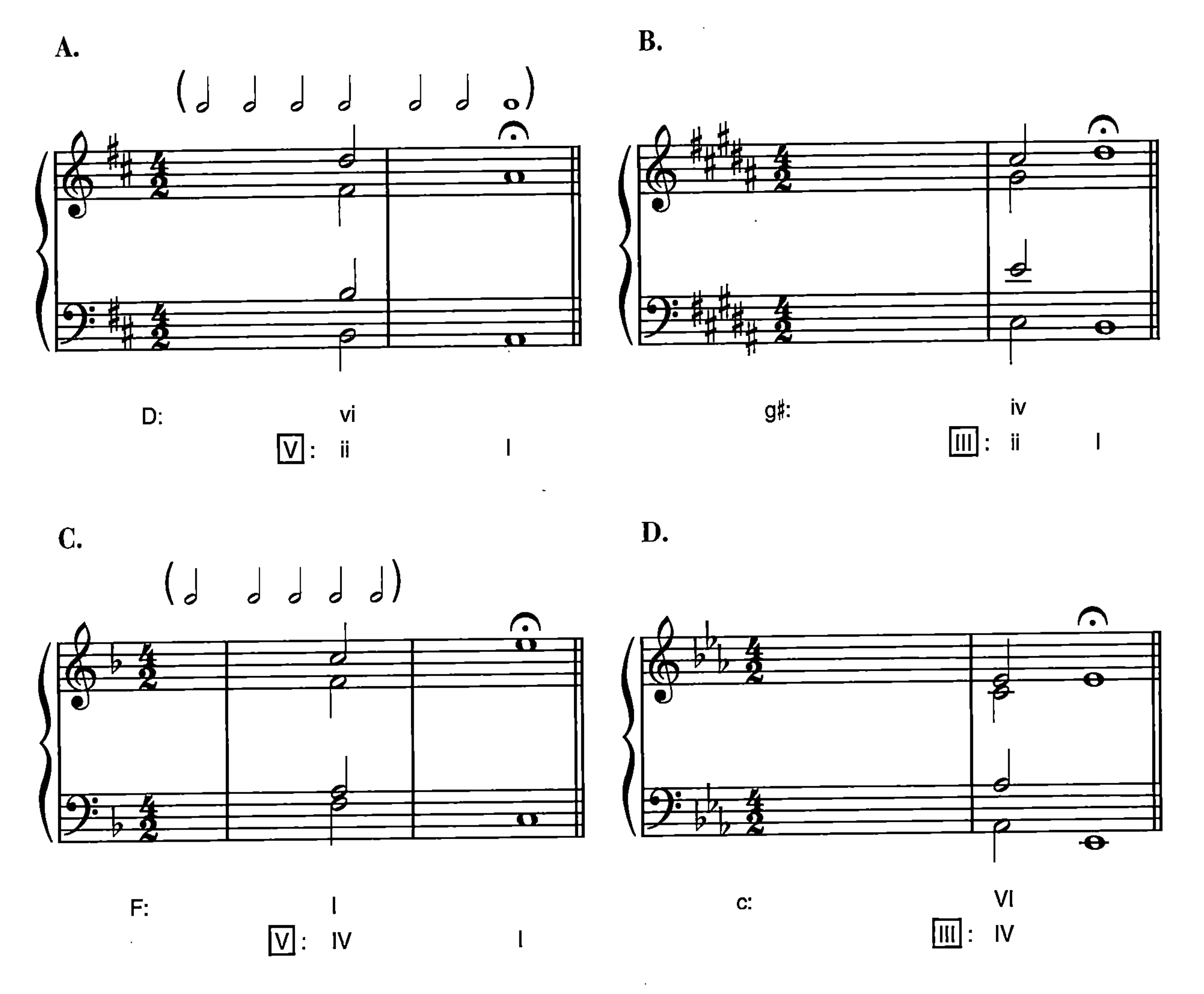

E.

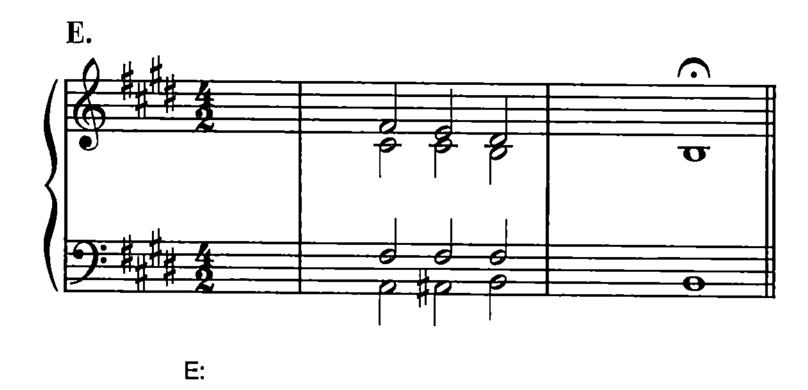

2 The following figured-bass exercises consist of two phrases each. The first phrase modulates to either [V] or [III], and the second phrase returns to the original key. Fill in the alto and tenor lines, and supply a roman-numeral analysis. Be sure to indicate the pivot chord in your analysis, if you actually need one to modulate. Exercise 2A is mostly unfigured.

A.

B.

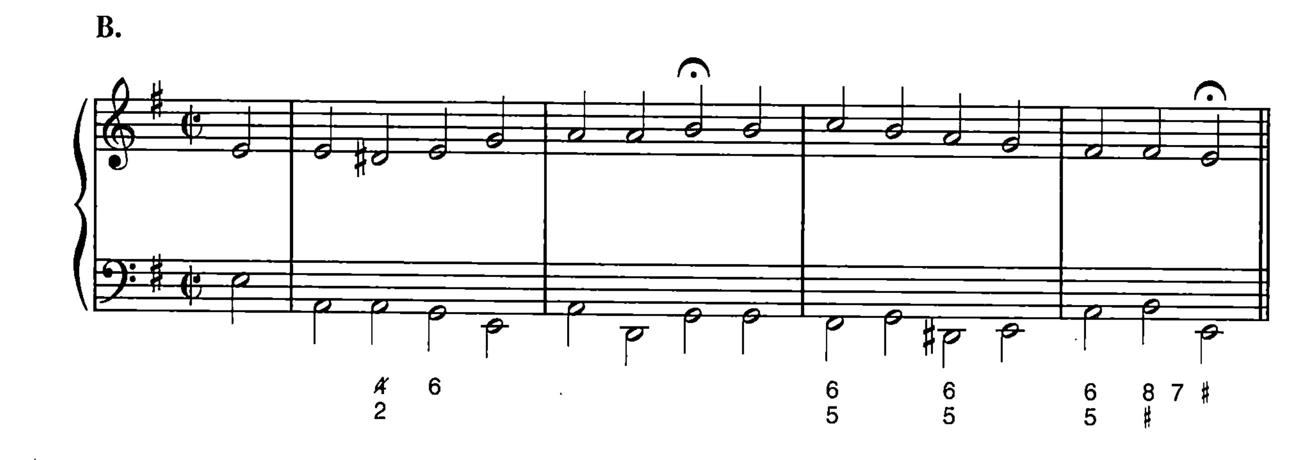

NAME ______________________________

C.

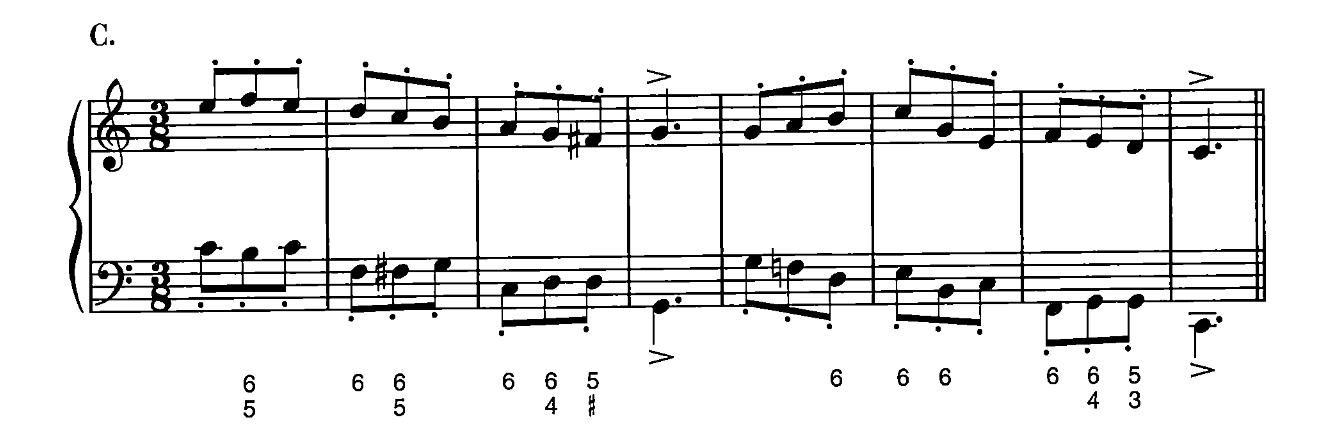

D.

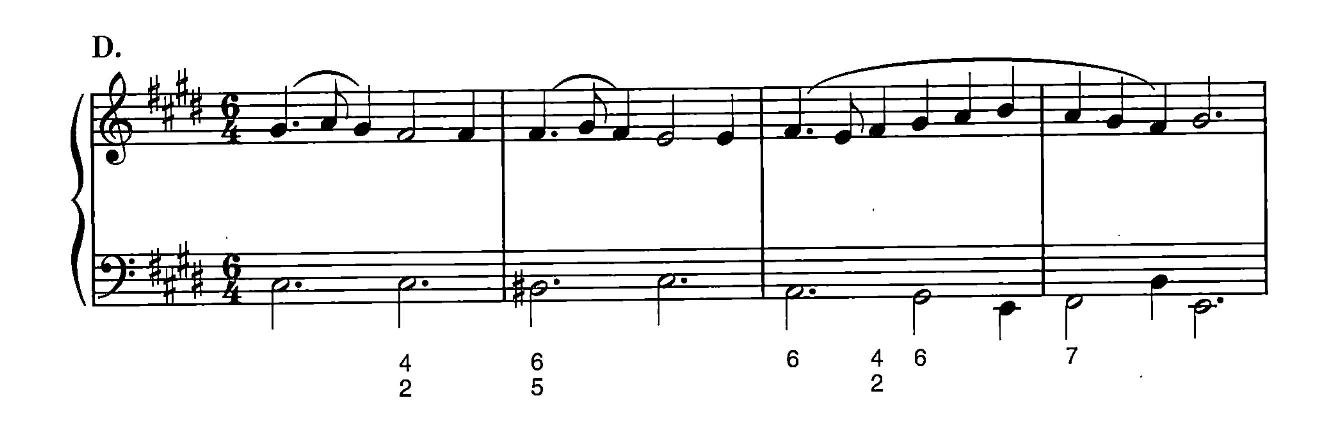

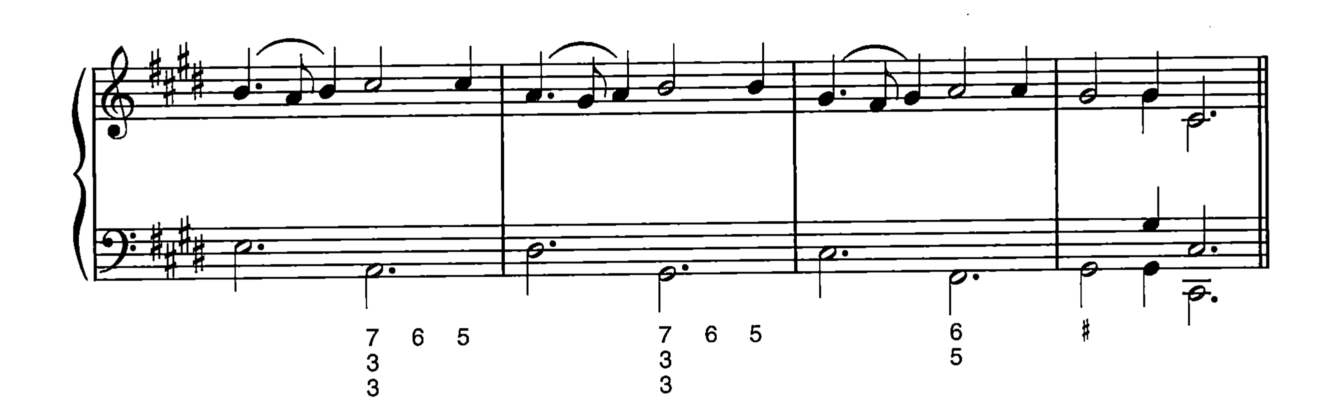

3 Harmonize one of the following melodies, which contain approximately the same number of harmonic changes (the second melody is incomplete). Determine where the tune cadences in a new key, and then sketch in an appropriate roman-numeral analysis. Write a bass line that complements the tune. You should set the chorale tune in four-voice texture; you may choose a different type of texture for the folk song.

A. "The Vicar of Bray"

B. "Sei gegrüsset, Jesu gütig"

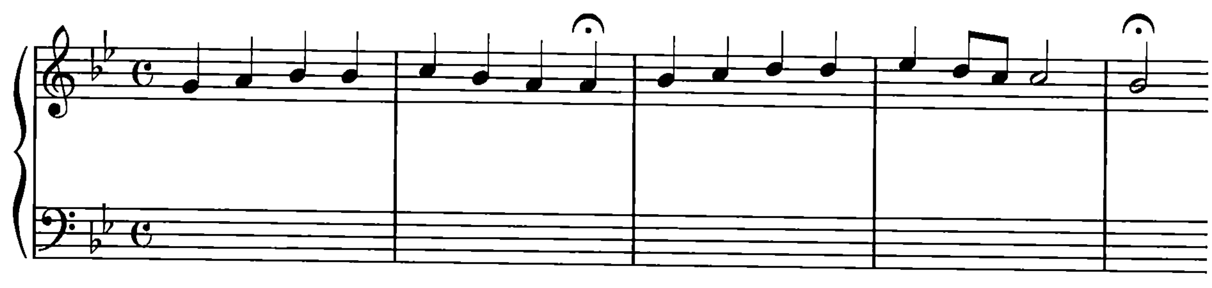

NAME ______________________________

4 Compose an original two-phrase period that modulates from a minor key to III and then returns to the original tonic. Strive for an interesting rhythmic setting. You might use the harmonic scheme of one of the preceding exercises as a model.

5 Excerpts from two Minuets are given below.

A. Make a voice-leading reduction of the Beethoven Trio and provide a roman-numeral analysis below the staff. Since this passage is in two-voice texture, you will have to infer the chords in a manner similar to unfigured-bass exercises. Use brackets above the staff to mark the extent of each phrase in this Trio. The D$\sharp^3$ in measure 5 indicates a V^6/vi. What happens tonally in measures 9–12? What rhythmic device does the slurring suggest in these four bars? The voice leading of the last phrase is a bit tricky; remember that since the harmonies change only once per measure, one soprano note per measure will suffice. Which ones make the most convincing line?

Beethoven: Minuet in G, WoO 11, No. 2, Trio

NAME ________________________________

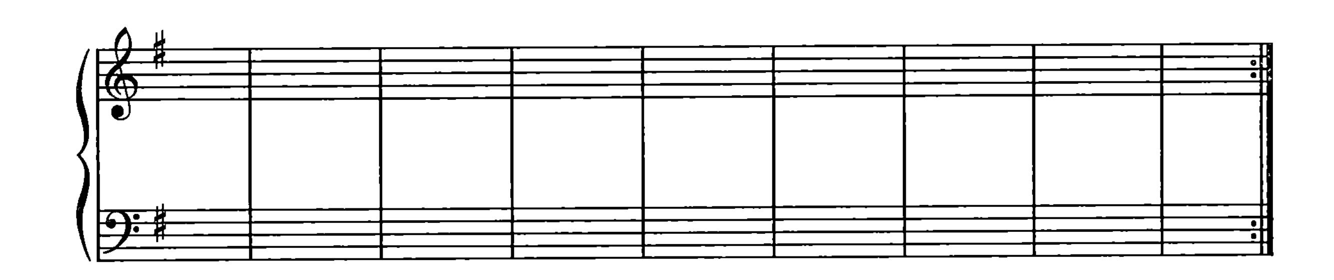

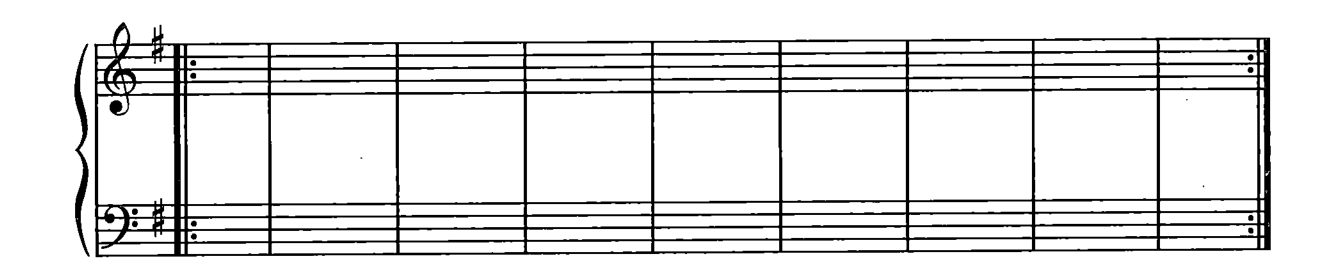

B. Provide a roman-numeral analysis of this piece. Why do you think the harmony in measure 12 is prolonged through measure 14?

BACH(?): MINUET IN D MINOR FROM *ANNA MAGDALENA BACH'S NOTEBOOK*

EXCURSION 1

Simple Forms

In this and the later Excursion chapter in this workbook, we suggest some pieces or movements for possible formal analysis. Many of these may be found in published anthologies of music.

One-Part

Bach: Prelude in C minor for Lute, BWV 999.
Chopin: Prelude in B minor, Op. 28, No. 6.

Binary or Two-Reprise

Simple Two-Part Form (AB)

Some folk songs consist of two well-defined sections: verse (A) and refrain or chorus (B); for example, see "My Bonnie Lies over the Ocean" and "My Old Kentucky Home." Many chorale tunes employ the form AAB, known as *bar form:* see "Christ lag in Todesbanden" and "O Haupt voll Blut und Wunden."

Baroque Two-Reprise Form

The dance movements in Suites or Partitas of this period are cast in two-reprise form. The shorter movements (Sarabande or Minuet) are usually less complex. Also consult the brief Minuets in Bach's *Notebook for Anna Magdalena Bach.*

Bach: Brandenburg Concerto No. 1 in F major, IV. This movement contains no less than four such sections. Do one or more employ rounded binary form?
Handel: *Water Music* or *Royal Fireworks Music.* Many of the interior dance movements employ two-reprise form.
Scarlatti: Sonata in D minor, L.413/K.9.

Classical Two-Reprise Form

Almost all of the Minuet and Trio sections of Classical Minuets employ rounded two-reprise form. So do many of the opening themes in theme-and-variations sets, as well as some of the initial refrains of rondos. Several interesting examples are listed below:

Diabelli: Waltz in C major (on which Beethoven based his famous Diabelli Variations for Piano, Op. 120).
Haydn: Symphony No. 104 in D major ("London"), II. See the opening section (mm. 1–37).
Mozart: Symphony No. 40 in G minor, III.
Beethoven: Piano Sonata in D major, Op. 28, III.

Ternary Form (ABA)

Examples of three-part form (ABA) may be found in slow movements of the Classical period, some songs of Schubert, shorter waltzes and mazurkas of Chopin, and certain of Brahms's Intermezzi.

Mozart: Piano Sonata in C major, K. 330, III. This movement also makes use of two-reprise form within each section.
Beethoven: Bagatelle in A major, Op. 33, No. 4. In what form is the initial section cast?
Schubert: "Pause" (from *Die Schöne Müllerin*) and "Auf dem Flusse" (from *Winterreise*). In both songs the last section is modified and/or extended.
Brahms: Intermezzo in A major, Op. 118, No. 2, and Intermezzo in E major, Op. 116, No. 6.
Chopin: Waltz in C♯ minor, Op. 64, No. 2, and Mazurka in A minor, Op. 7, No. 2. Contrast the form of the initial sections of these works.

Variation Form

Continuous Variations of the Baroque period

Purcell: Dido's Lament (from *Dido and Aeneas*).
Bach: Chaconne (Partita for Solo Violin in D minor).
Buxtehude: Passacaglia for Organ in D minor.
Handel: Chaconne in G major (*Trois Leçons*).

Theme-and-Variations Sets of the Classical Period

Mozart: Piano Sonata in A major, K. 331, I.
Mozart: Piano Sonata in D major, K. 284, III.
Beethoven: Variations on "God Save the King," WoO 78.
Beethoven: Piano Sonata in A♭ major, Op. 26, I.
Beethoven: Symphony No. 5 in C minor, II. This freer handling of variation form uses two themes, of which only the first is varied.

Rondo Form

Mozart: Piano Sonata in C major, K. 545, III. A very condensed rondo form.
Haydn: Symphony No. 102 in D major ("Clock"), IV. This movement contains some interesting modifications of the initial refrain.
Beethoven:Piano Sonata in G major, Op. 49, No. 2, II.
Beethoven: String Quartet in C minor, Op. 18, No. 4, IV.

These Beethoven movements are almost "textbook" examples of the two major types of rondo.

NOTES

NOTES

NAME ____________________

CHAPTER 20

Harmonic Sequences I

ROOT MOVEMENT BY 2ND AND 3RD

1 Continue the following passages in sequential fashion. Note that the first two employ three-voice texture. Identify the basic type of sequence illustrated in each passage: motion by descending or ascending 2nd or 3rd. How are parallels avoided in Exercise 1B?

A.

B.

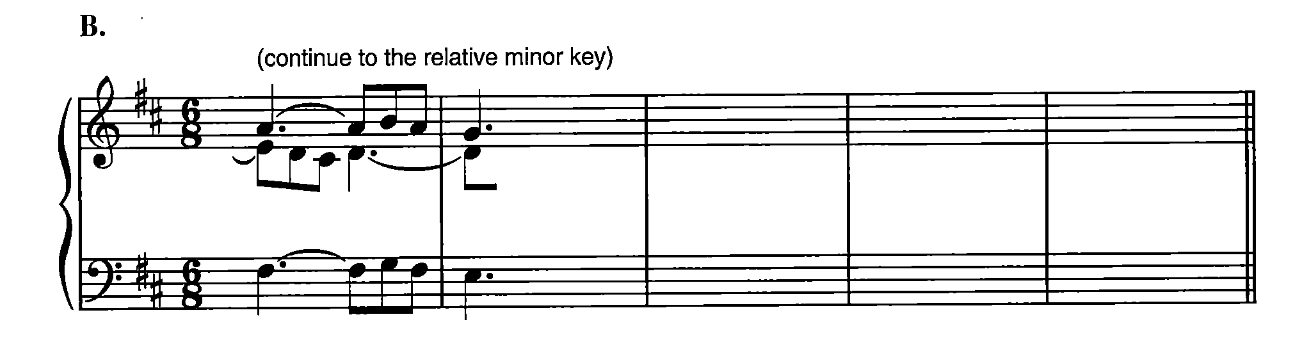

C.

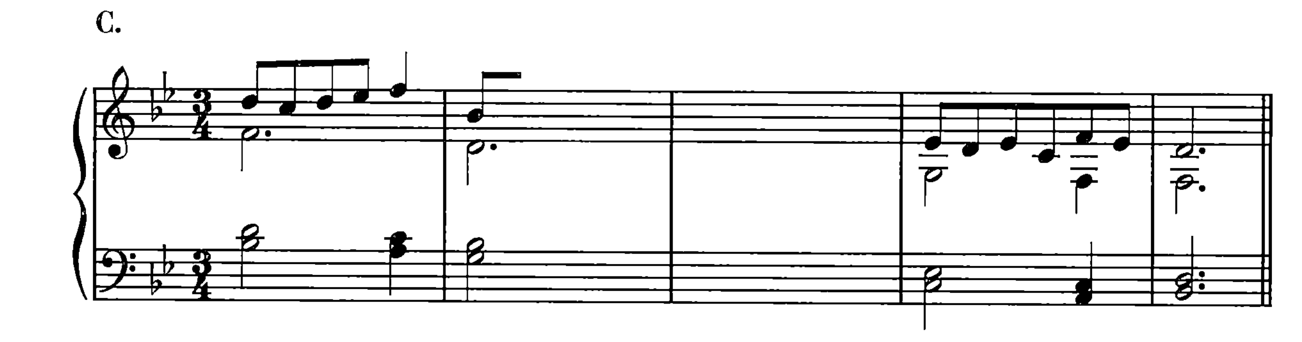

2 Realize the figured-bass exercises below. In your roman-numeral analysis, use only figured-bass symbols for those sections that are sequential. Be careful not to create parallels in Exercise 2A.

A.

B.

NAME ________________________________

C.

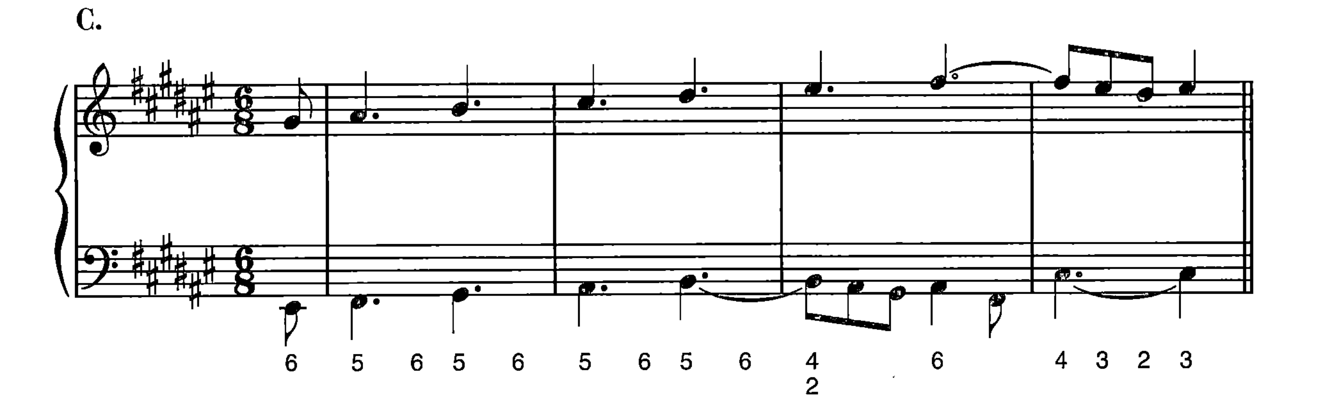

D.

3 Compose a harmonic setting for the following two melodies, using an appropriate sequence featuring root movement by 2nd or 3rd for the sections marked with brackets. You may use three-voice texture for the first tune.

A.

B.

4 Analyze the following excerpts, denoting the type of sequence employed in each. Make a voice-leading reduction of each excerpt on the empty staves provided. You may wish to refer to similar models found in Chapter 20 of the text.

A. While making your reduction, pay particular attention to the linear motion on the first beat of each measure.

BACH: BRANDENBURG CONCERTO NO. 1 IN F MAJOR, III

B. What function do the chords on the second beats of the first three measures perform?

BEETHOVEN: SYMPHONY NO. 5 IN C MINOR, I

C. What dissonances in measures 2 and 3 momentarily disguise the soprano motion?

GLUCK: *ALCESTE*, ACT III

D. This passage employs a two-voice texture. In order to show the alternating root position and first-inversion chords, it is necessary to imply a third voice, which fills out the chordal harmony; see the opening of the analysis.

SCARLATTI: SONATA IN F MAJOR, K. 518

E. Carefully observe the stemming in the upper voices of the first three measures below. In what unusual way does this passage avoid parallel 5ths?

CORELLI: CORRENTE FROM CONCERTO GROSSO NO. 10 IN C MAJOR

NAME ______________________________

CHAPTER 21

The Leading-Tone Seventh Chord

1 For the following progressions, only the key and bass line are provided. Complete the chords in four-voice texture, employing a vii^{o7} (or vii^{ø7}) in various inversions on those chords marked with an arrow. Indicate how the approach to and resolution of the chordal 7th is handled (by suspension, neighboring tone, etc.). Be careful not to create similar 5ths, except those that occur in $\frac{6}{5}$ inversion.

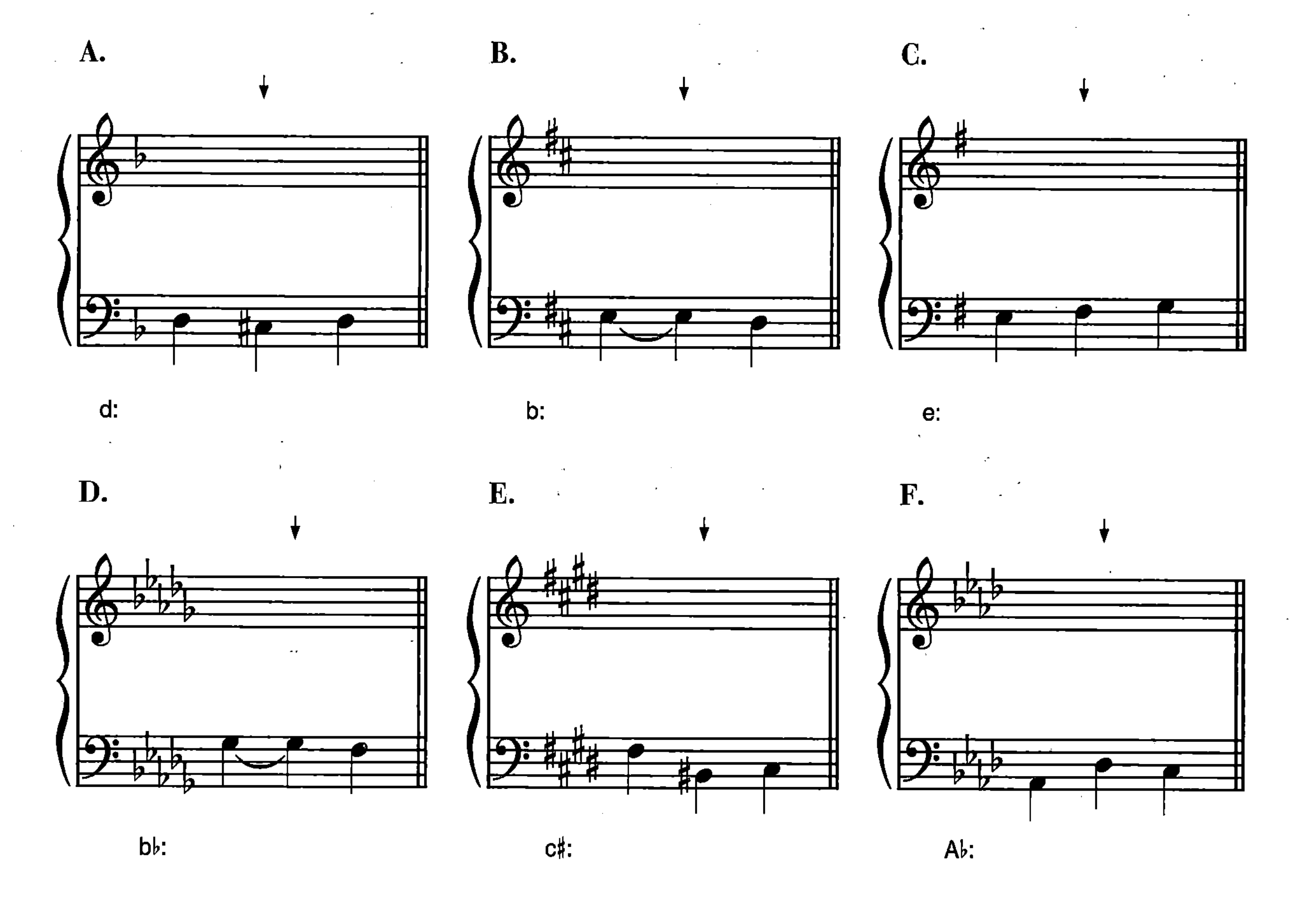

2 Realize the following figured-bass exercises and provide a roman-numeral analysis for each passage. Explain how the chordal 7th of each leading-tone seventh chord is handled. Exercise 2B is unfigured; try to use two vii°7's in different positions.

A.

B.

C.

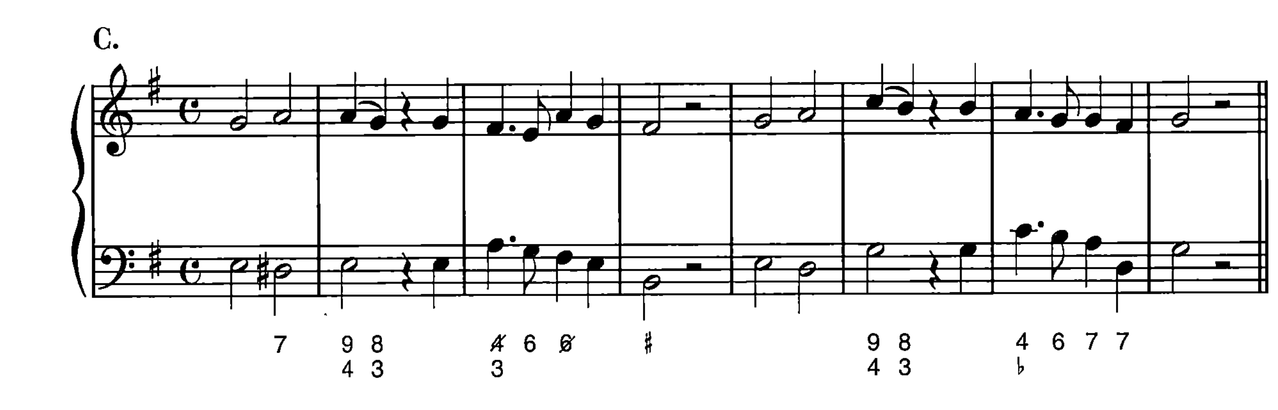

D.

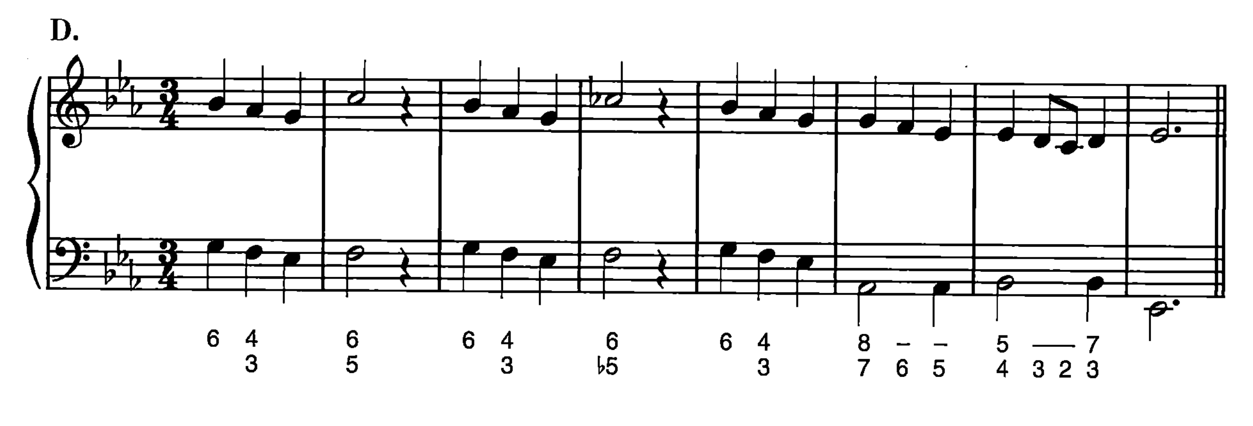

E.

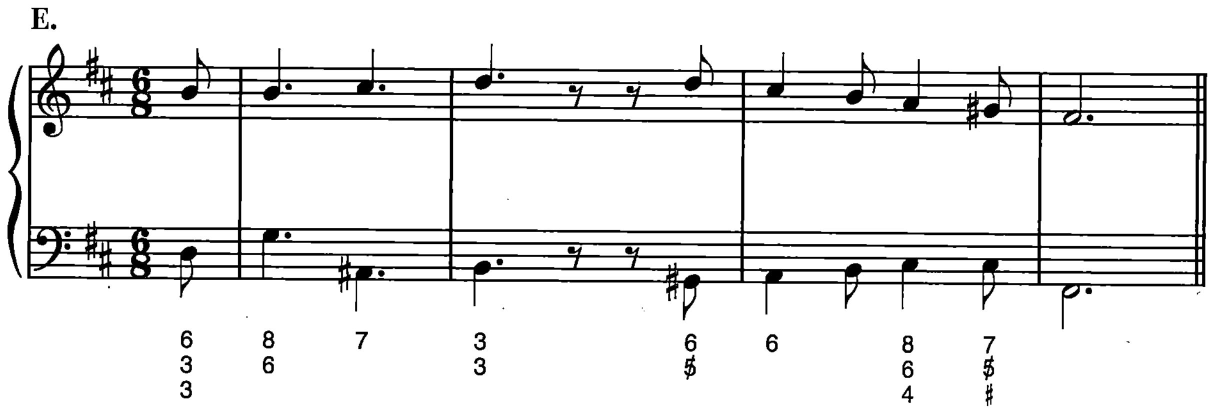

3 Harmonize the two melodies below, using a vii°7 or one of its inversions at those points marked with an arrow. Notice that the second tune appears in the bass voice!

A.

B.

4 Make voice-leading reductions of the following excerpts, and provide a roman-numeral analysis for each.

A. Why do you think measures 4, 6, and 8 are not strong cadential points in the piece? Why do you think the last two measures are?

GLUCK: "THROUGH THIS GROVE" FROM *ORFEO ED EURIDICE*, ACT I

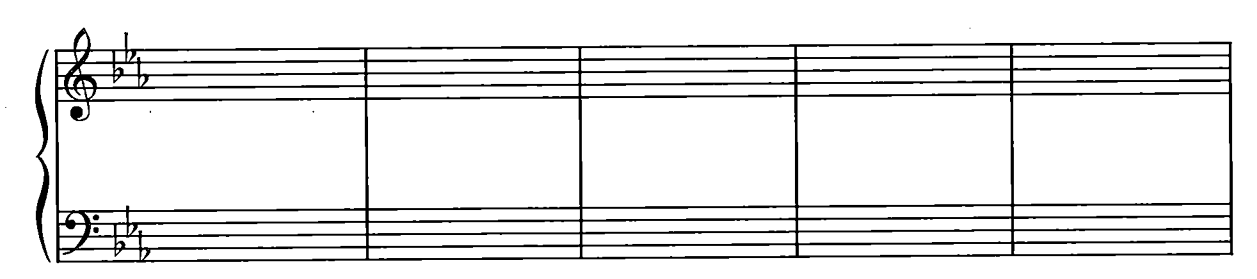

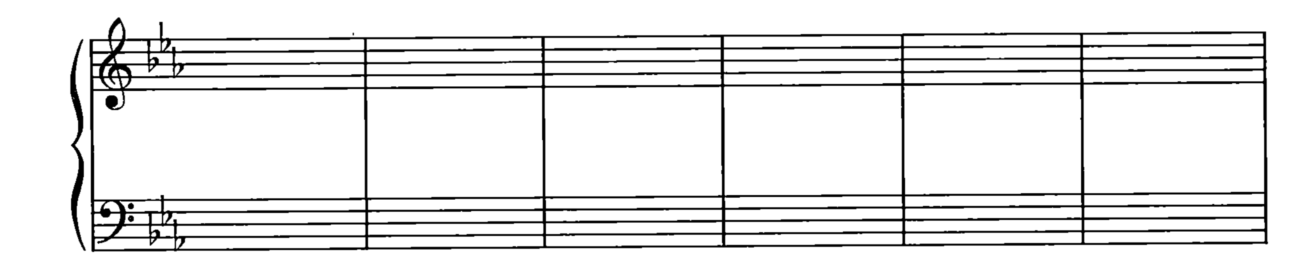

B. What is peculiar about the rhythmic/metric setting of the vii°7's in this passage?

BRAHMS: BALLADE IN B MAJOR, OP. 10, NO. 4

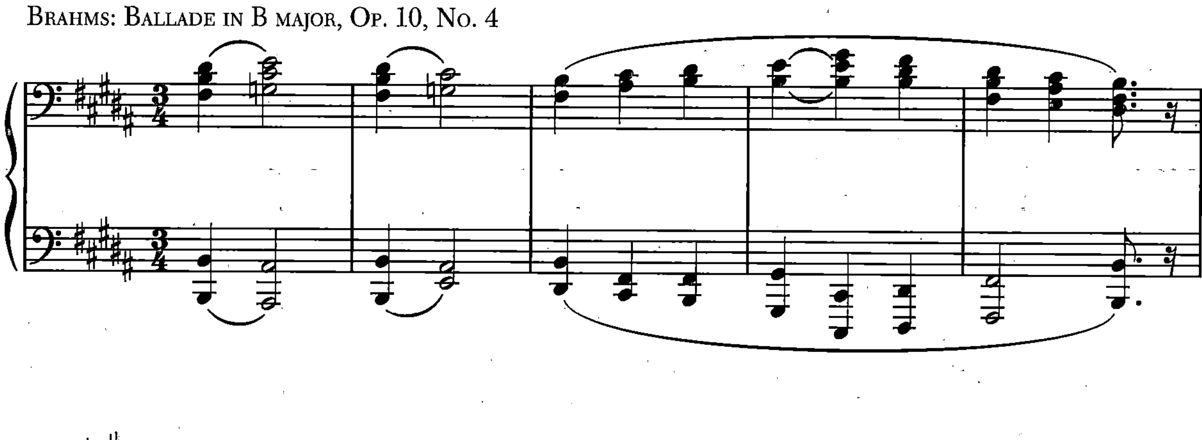

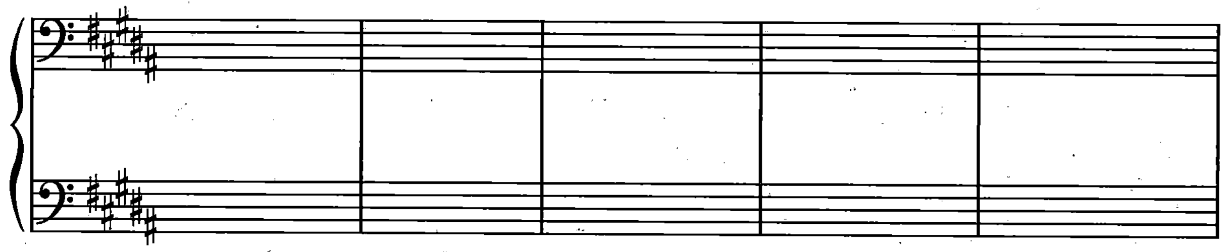

C. The key signature of this quotation is misleading; the key is actually E♭ minor! What type of nonharmonic tone occurs in the first three measures? Can you find a diminished seventh chord that is *not* built on the leading tone?

FRANCK: PRELUDE, CHORALE, AND FUGUE FOR PIANO

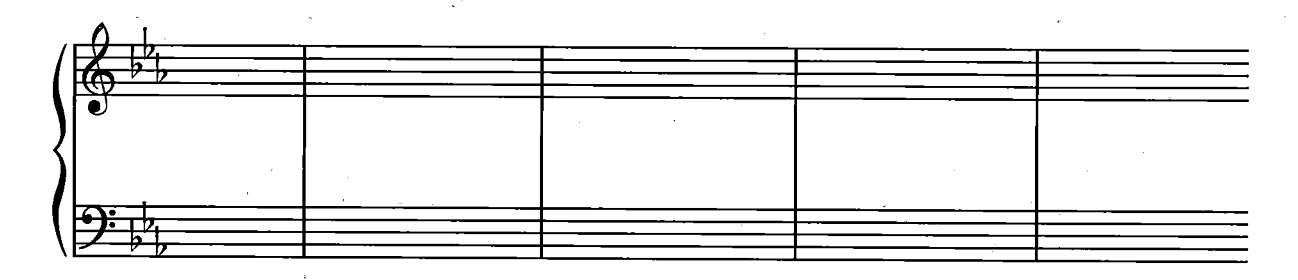

D. This very interesting passage contains numerous accented nonharmonic tones as well as a double meter signature. What do you think is the key center? What chord is being prolonged through the first two beats of measure 4? The final close is a typical Wagnerian deceptive cadence.

WAGNER: *DIE WALKÜRE*, ACT II, SCENE 3

NAME ______________________

CHAPTER 22

Harmonic Sequences II

ROOT MOVEMENT BY 5TH

1 Realize the opening of each figured-bass passage below, and then continue in like sequential manner until the cadence. Brackets signify a full sequential statement, half brackets a partial one. Indicate the key of the passage and label the first and last harmonies with roman numerals. What is the underlying root movement in each example?

A.

B.

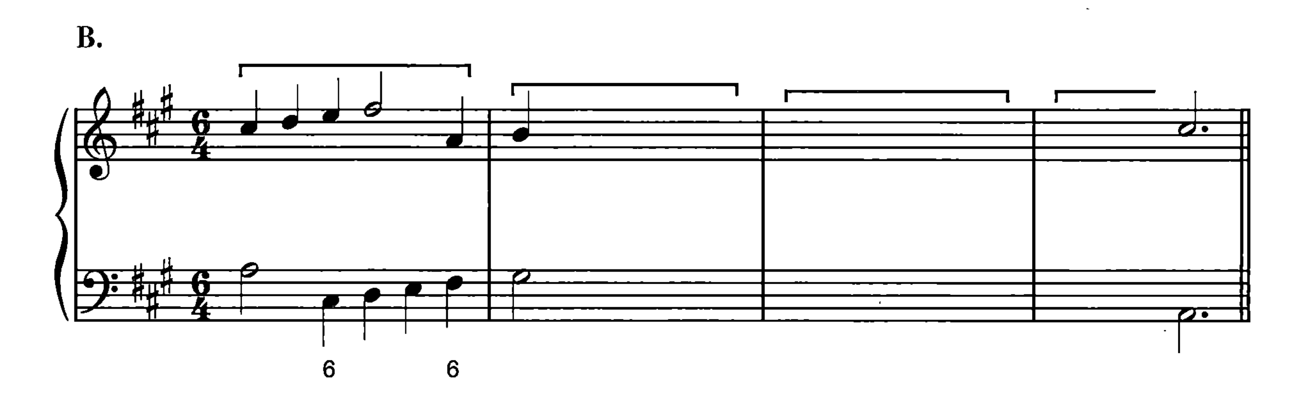

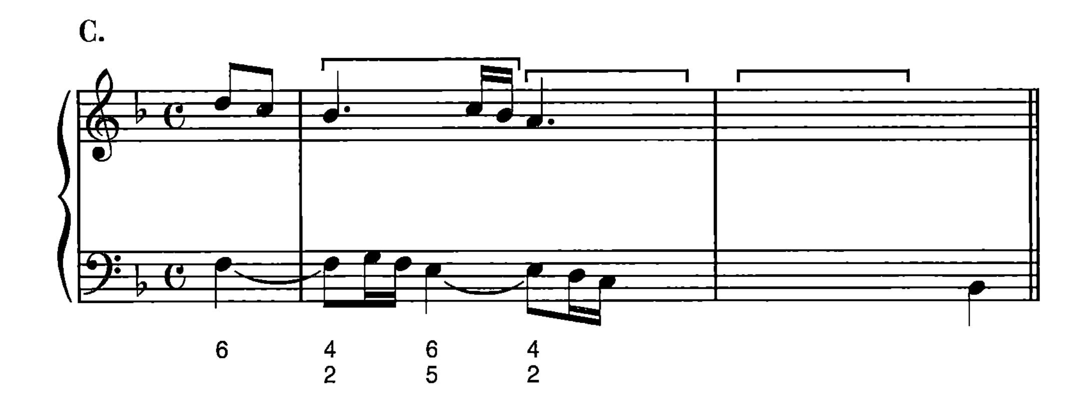

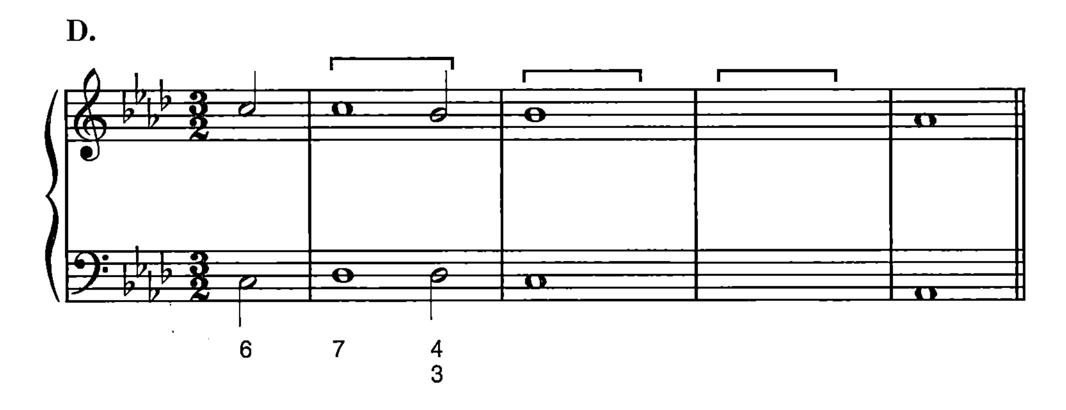

2 On the staves after each passage, elaborate each of the three-voice sequences. Be careful that you do not create parallels. Your figuration should remain exact until the sequential motion ends. Choose your own meter and motivic ideas. Identify each type of sequence with figured-bass symbols. One possible elaboration for Exercise 2A is given in the model.

NAME ___________________________________

B.

3 Harmonize the two melodies below, employing sequences by descending 5th. Try to incorporate some chordal inversions by means of your bass lines.

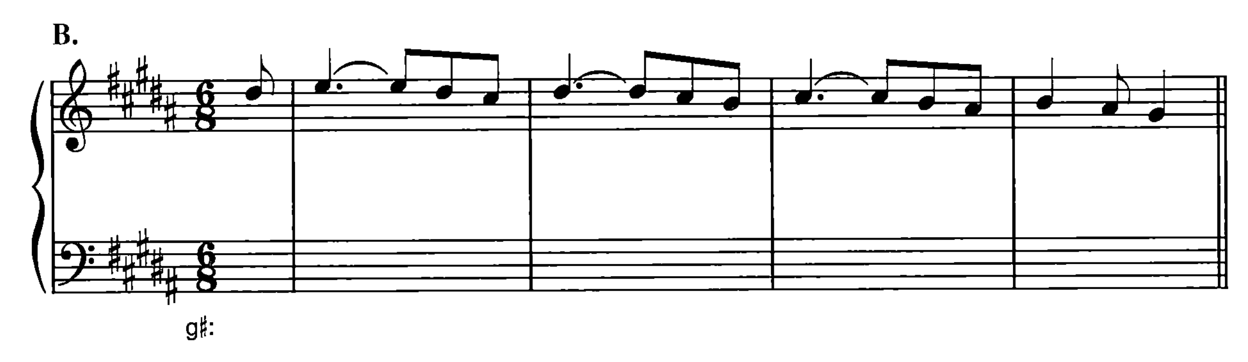

NAME ______________________________

4 The following excerpts display various types of diatonic sequences. Make voice-leading reductions of each excerpt in the space provided, and then below the staves label the essential harmonies with roman numerals and the sequential motion with figured-bass symbols.

A. In your reduction, show how this sequence is derived from a series of stepwise 7–6 suspensions.

CORELLI: ALLEMANDE FROM TRIO SONATA DA CAMERA IN A MAJOR, OP. 4, NO. 5

B. The theme (or subject) of this fugue was composed in such a way as to suggest ornamented suspension figures. With this in mind, in which voice part do you think it appears in this excerpt? Remember that the second part of a tied note in a suspension is always dissonant.

BACH(?): FUGUE IN G MAJOR FROM *EIGHT LITTLE PRELUDES AND FUGUES* FOR ORGAN

C. What type of nonharmonic tone elaborates this sequence?

HANDEL: "SURELY HE HAS BORNE OUR GRIEF" FROM *MESSIAH*

D. Note the pun in the title! How is the $\frac{5}{4}$ meter grouped?

PAUL DESMOND: "TAKE FIVE"

E. Bracket the sequential patterning in the eighth notes. What meter does this suggest?

BACH: THREE-PART INVENTION IN E MAJOR

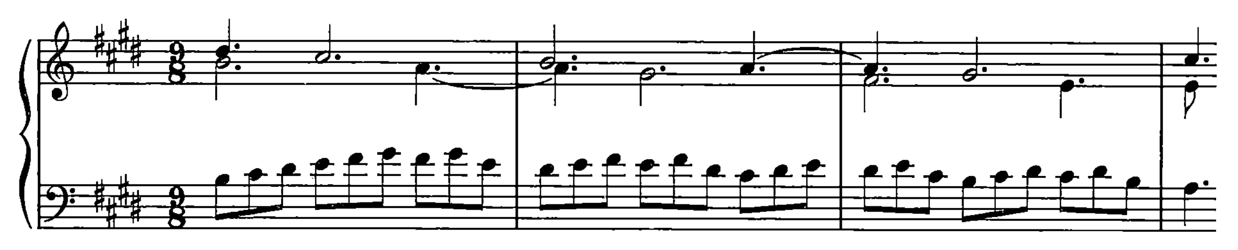

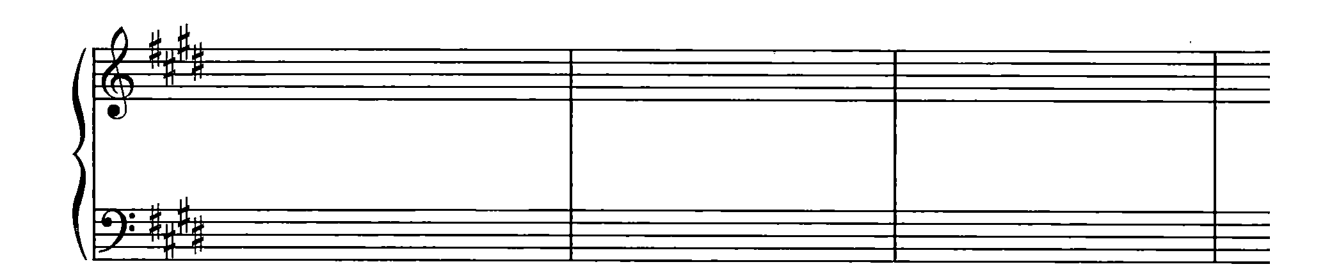

5 An excerpt from a well-known operatic aria is given below. Considering the failing state of Mimi's health, the basic direction of the melody seems appropriate. What voice part in the accompaniment doubles the vocal line at the octave? Although this passage is not strictly sequential, it does contain some brief sequential passages. Analyze it with roman numerals, and bracket the portion that suggests a progression by descending 5ths.

PUCCINI: "SONO ANDATI?" FROM *LA BOHÈME,* ACT IV

. . . Fingevo di dormire
perchè volli con te sola restare.
Ho tante cose che ti voglio dire
o una sola, ma grande come il mare, . . .

. . . I pretended to sleep
because I wanted to remain alone with you.
I have so many things I want to say to you
or just one, but as big as the sea, . . .

NAME ______________________________

CHAPTER 23

Two Analysis Projects

(textbook chapter title: Analytical Comments on a Menuetto and Trio by Beethoven)

Two extensive passages are quoted below. The first is the opening section of a Mozart minuet; the second is a complete prelude from the Baroque period. Choose one of these and make a voice-leading reduction, provide a roman-numeral analysis, and comment on the themes, motives, phrasing, texture, and so forth. In addition, address the points that are stated in the directions for each excerpt.

1 The Minuet section of Mozart's Symphony No. 36 in D major ("Linz") is cast in a two-reprise form. However, there are several significant deviations from the usual design (diagrammed in Chapter 23 of the text), which incorporate tonal changes, thematic material, and phrase grouping. Identify some of these.

MOZART: SYMPHONY NO. 36 IN C MAJOR ("LINZ"), III, MINUET

tr
tr
11
(f)
p
19
tr

NAME ____________________

____

____

____

____

____

2 The C-major Prelude that opens the first volume of Bach's monumental *Well-Tempered Clavier* is actually a later revision of the original version. (See the Kalmus edition for Hans Bischoff's comments on the earlier versions.) The persistent sixteenth-note patterns shown here in the initial four measures continue throughout the rest of the piece.

There are several instances of diminished and major-minor seventh chords that function as secondary dominants to scale degrees other than the tonic. These are labeled in the score. The A^5 and G^5 in measures 5 and 7 momentarily cover the stepwise soprano voice leading. Compare the tonal motion within measures 7–11 with that within measures 15–19. How are the two chords in measures 21 and 24 diatonically connected? What is the function of the passage in measures 24–31? Why do you think Bach resolves this section to a tonic triad with $\flat\hat{7}$ (or V^7/IV)? Before you answer this last question, try playing this piece, resolving the chord in measure 31 to the last chord, thereby skipping measures 32–33 altogether. After examining your analysis, you might wish to consult the voice-leading graph of this prelude by Heinrich Schenker, found in his *Five Graphic Analyses* (New York: Dover, 1969).

BACH: PRELUDE IN C MAJOR FROM *WELL-TEMPERED CLAVIER*, BOOK I

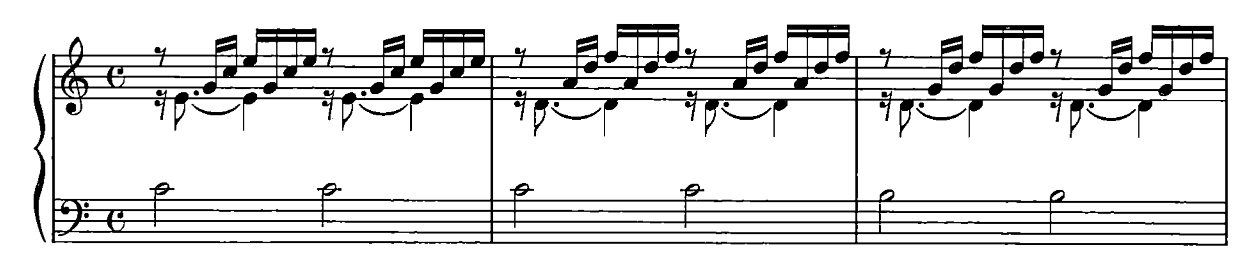

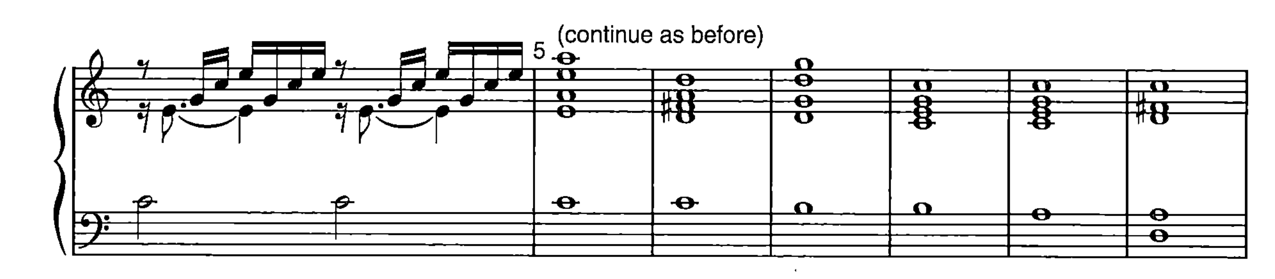

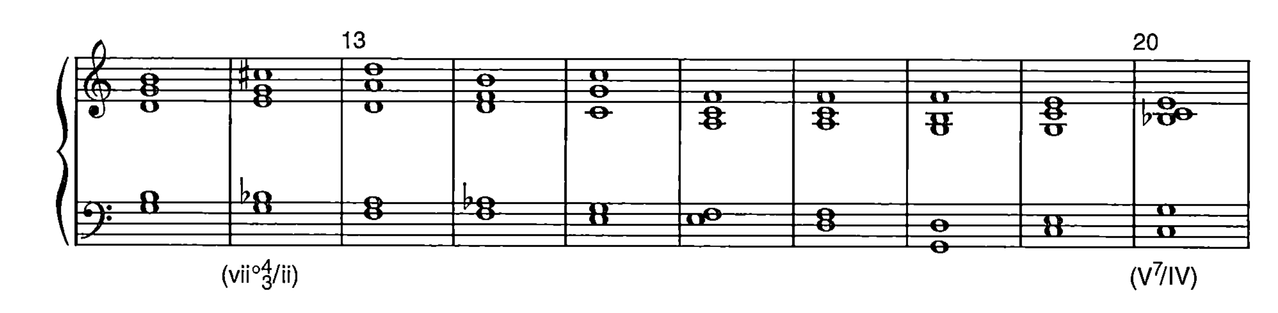

NAME ______________________________

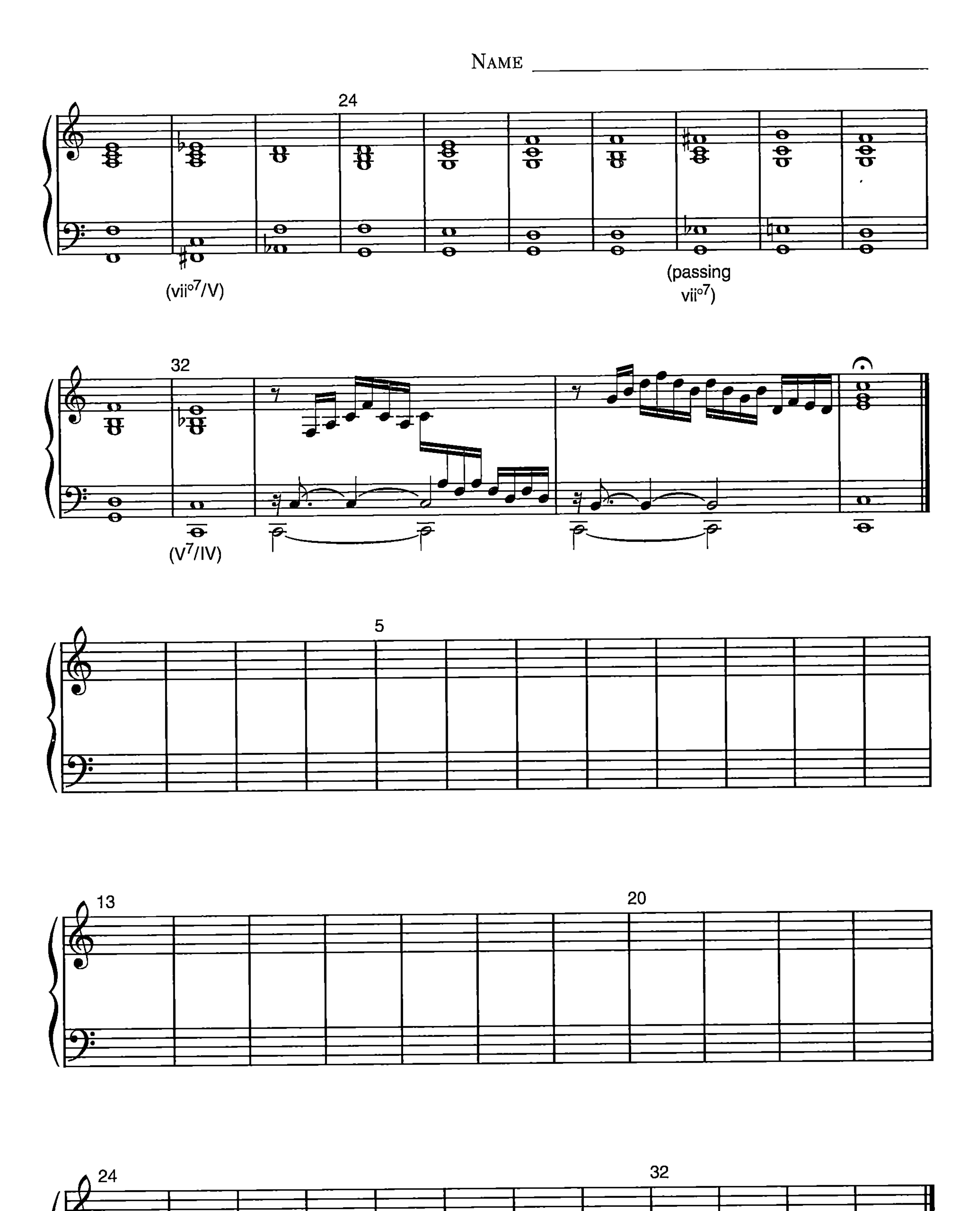

PART THREE

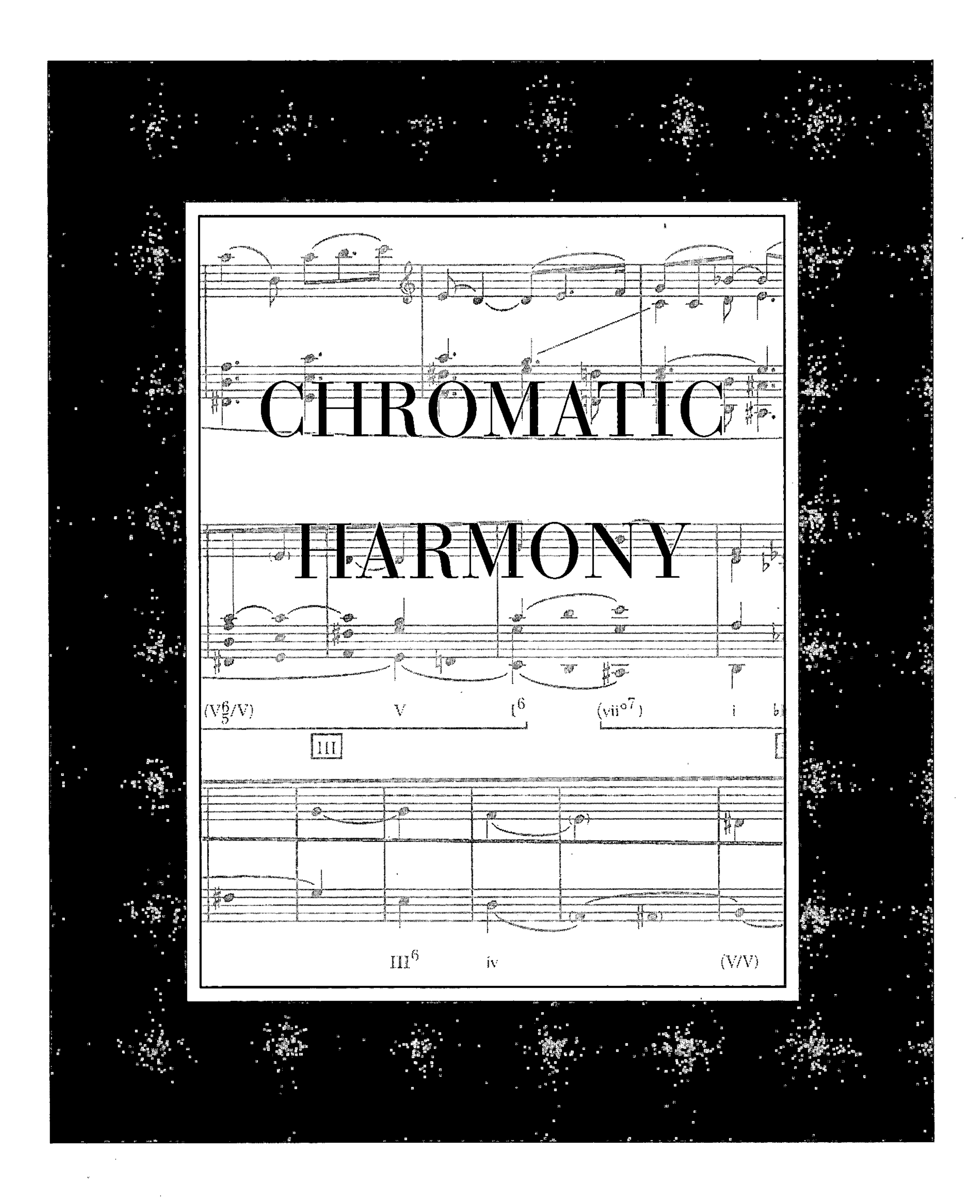

CHROMATIC HARMONY

NAME ______________________________

CHAPTER 24

Review of Diatonic Harmony

(textbook chapter title: Introduction to Chromatic Harmony)

Since this chapter does not require any student assignments, the following exercises review the material in Chapters 9–23.

1 Compose original phrases in four-voice texture that illustrate the chords or devices listed below. You may use chorale style or attempt more rhythmically varied settings. Be sure to stay within the indicated keys.

A. Use a V^4_3 chord and a cadential $^6_4$ in A major.
B. Use a $ii^{ø4}_3$ chord and a 4–3 suspension in F minor.
C. Use a vi chord and modulate to $\boxed{V}$ from G major.
D. Use a vii^{o4}_3 and end with a Phrygian cadence in B minor.
E. Use a voice exchange and a passing $^6_4$ in D♭ major.

2 The outer voices of the first section of a Mozart minuet for wind octet are quoted below. How are the two parts related? Is this relation strict throughout? To what key does this section modulate? Would you consider this key change a normal one for a first reprise in the minor mode?

Using the figured-bass symbols that are supplied, add two middle parts to complete a four-voice texture. Since your tenor voice must be rather high in some places, you may wish to write it in the treble clef. Take care with your partwriting, as there are several tricky spots. Supply a roman-numeral analysis.

MOZART: WIND SERENADE IN C MINOR, K.388, III

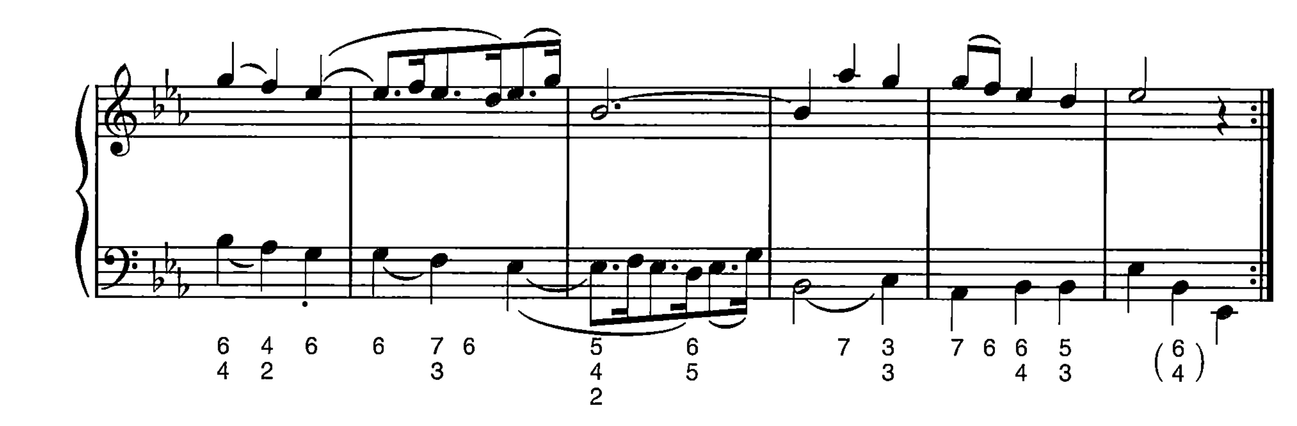

3 The passage that immediately follows the first reprise of the minuet given in Exercise 2 is quoted below. Make a voice-leading reduction of this passage, omitting all unison doublings. What type of sequence is exemplified here? What melodic dissonance does Mozart primarily exploit?

MOZART: WIND SERENADE IN C MINOR, K.388, III

NAME ______________________________

CHAPTER 25

Tonicization and Modulation II

SECONDARY DOMINANTS

1 A series of diatonic triads in various keys is given below. Precede each one with a secondary dominant (V, vii^{o6}, V^7, vii^{o7}, or their inversions) in four-voice texture. Use a different type of applied chord and inversion for each triad. Be sure to approach and resolve any chordal 7th correctly.

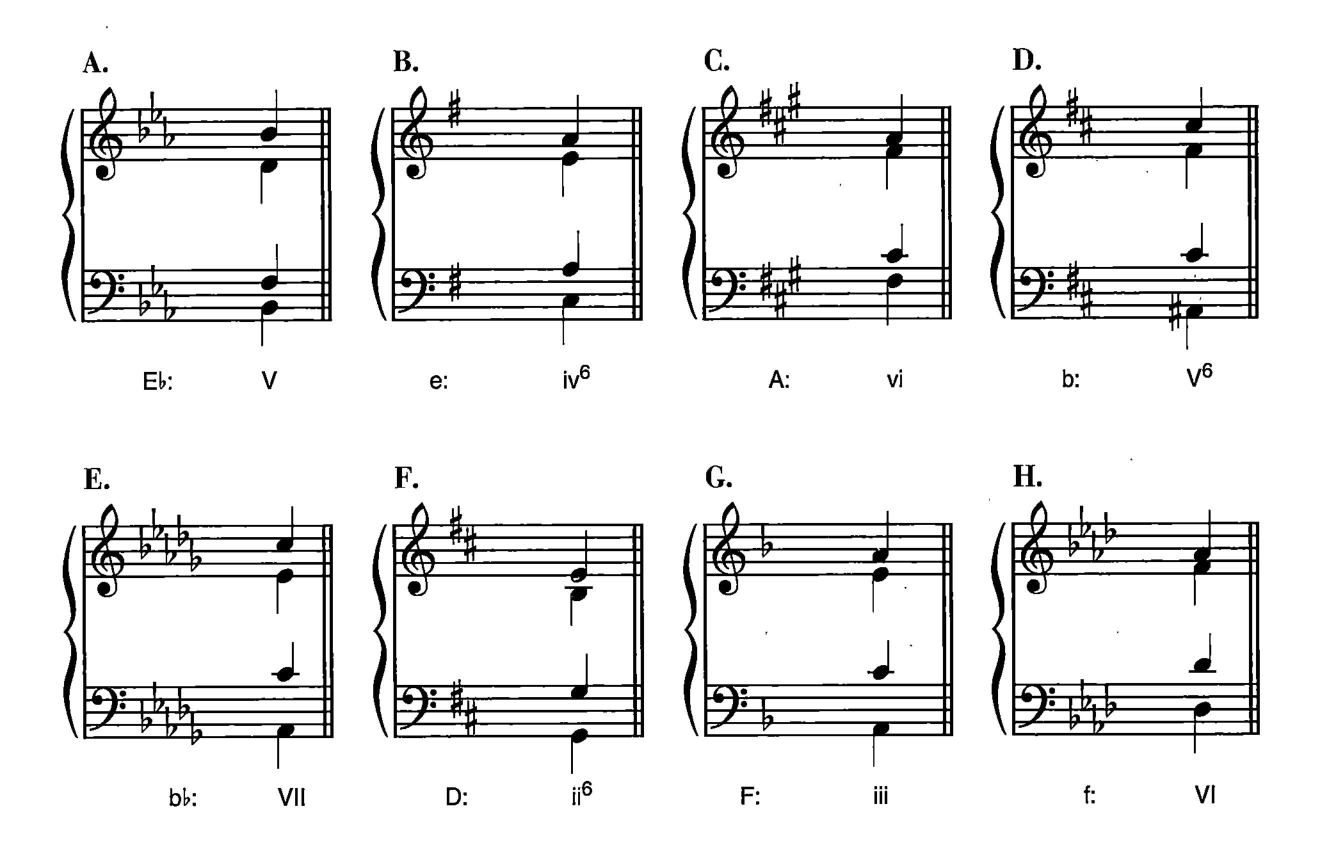

2 Realize the following figured-bass exercises, and supply a roman-numeral analysis for each. Exercise 2A is unfigured. In most cases, the applied chords will function as embellishing harmonies, but sometimes they may substitute for an essential pre-dominant. Bracket any passage that displays sequential movement.

The figured-bass exercises in Parts Three and Four will become progressively more chromatic. Always remember to realize the figures in terms of the key signature rather than the transient tonality you might be in at a given moment.

A.

B.

C.

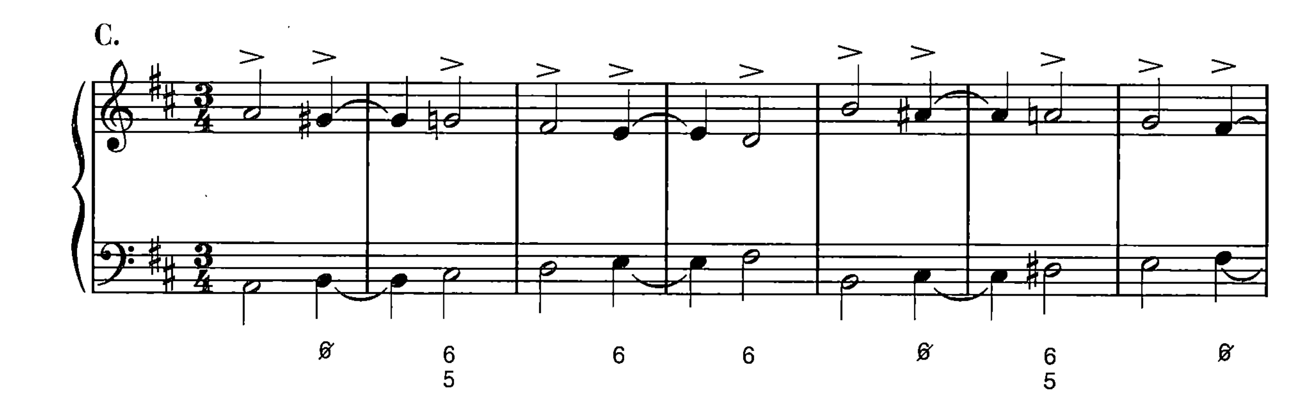

NAME ______________________________

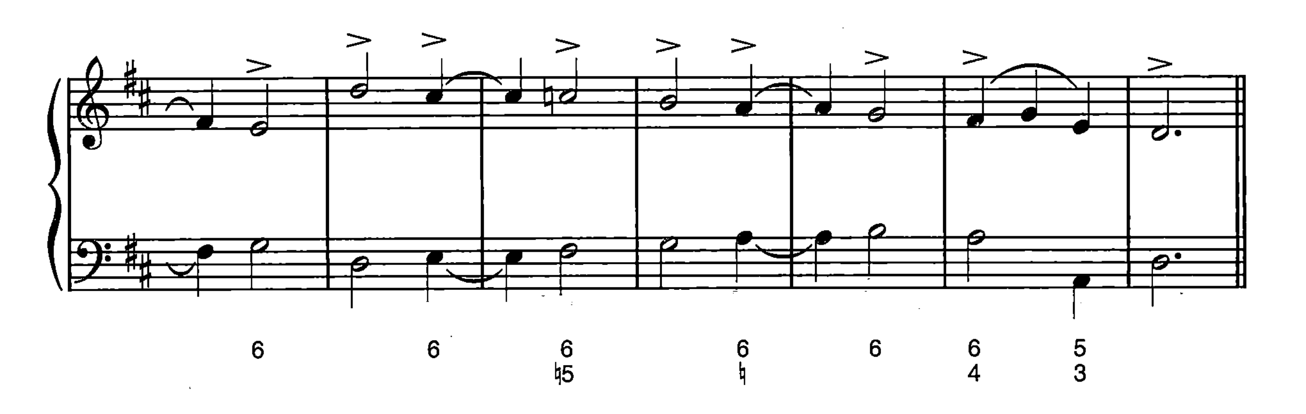

D.

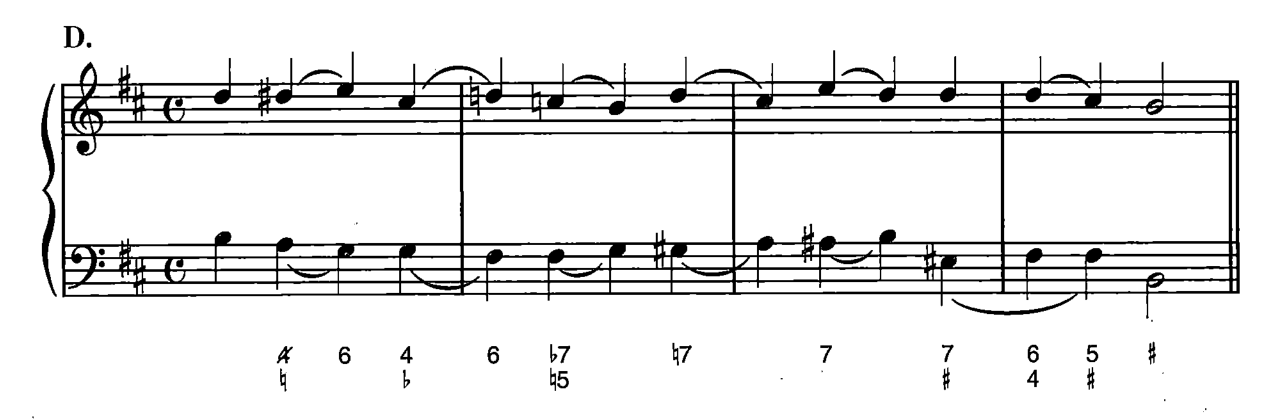

E.

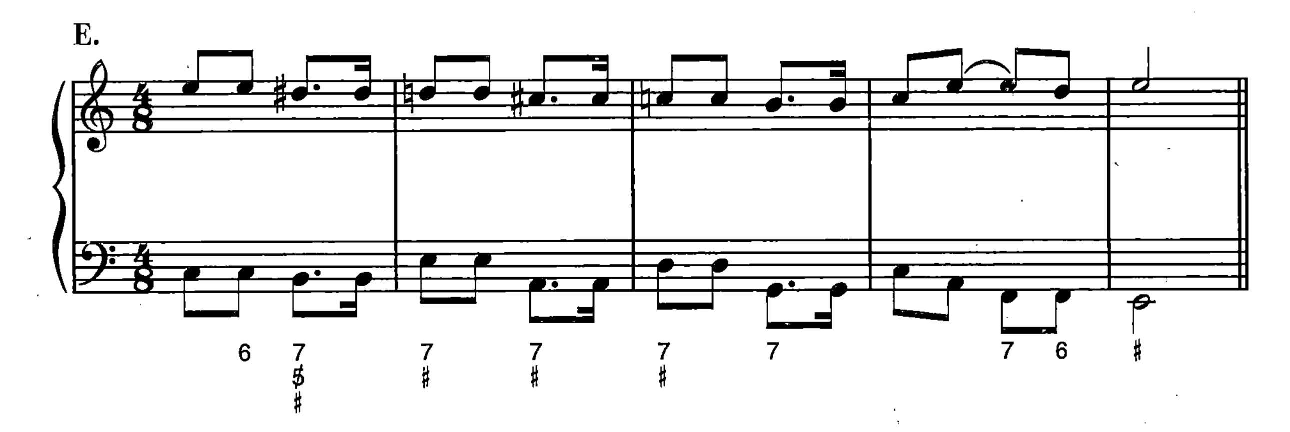

3 Harmonize the following two melodies, using a secondary dominant on those notes marked with an arrow. Your finished version should include a tonicization of the five basic diatonic triads other than the tonic in the given mode.

A.

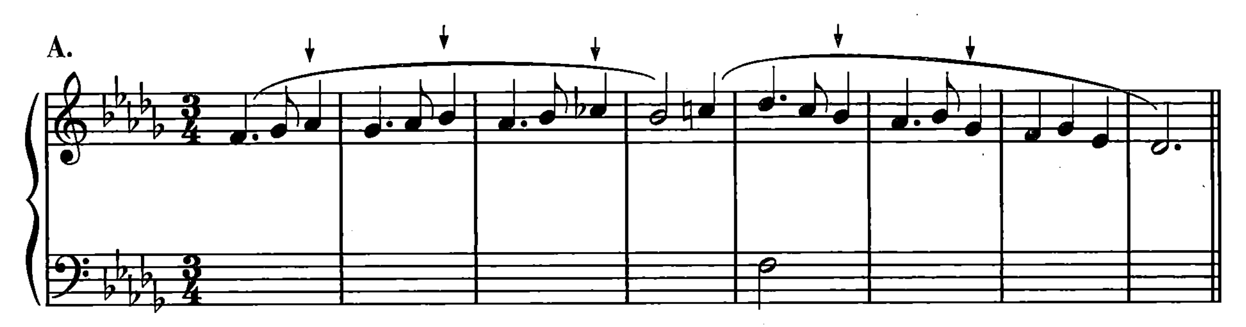

B.

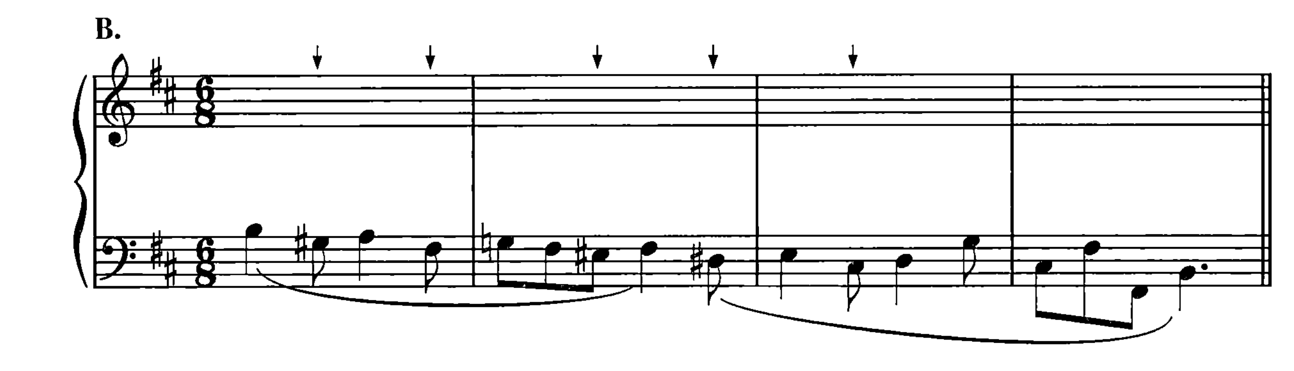

4 Make a voice-leading reduction of the following excerpts and then supply a roman-numeral analysis of each.

A. What chord is tonicized in measures 2–3?

Franck: "Panis Angelicus" (text omitted)

B. In your reduction, bracket any melodic or harmonic sequences.

VERDI: QUARTET FROM *RIGOLETTO*, ACT III (VOCAL PARTS OMITTED)

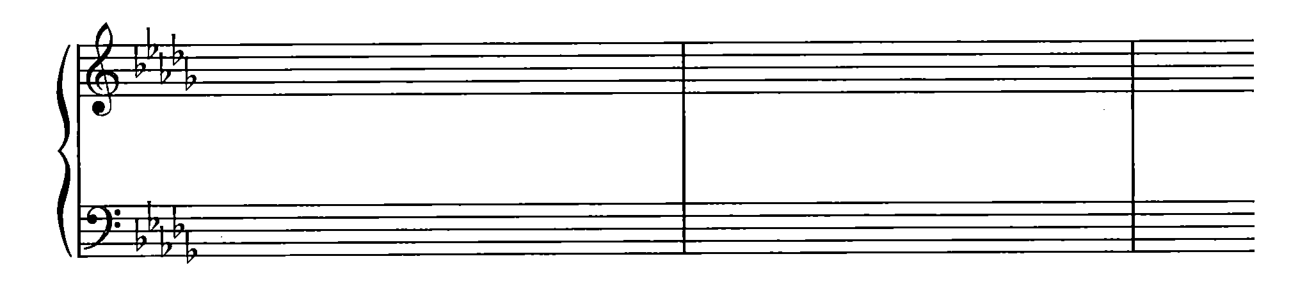

C. How do the applied chords resolve in this excerpt?

LISZT: *LIEBESTRAUM* NO. 3 IN A♭ MAJOR

D. Toward what new key are the final five measures directed? (The last chord does not function as the tonic in the new key.)

HAYDN: SYMPHONY NO. 102 IN D MAJOR ("CLOCK"), IV (SIMPLIFIED)

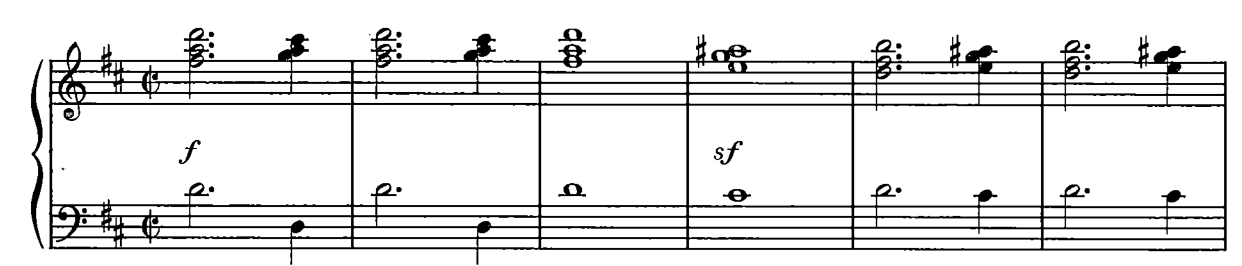

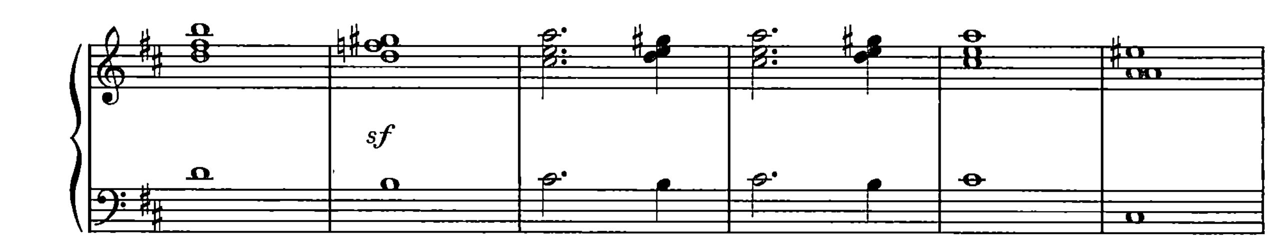

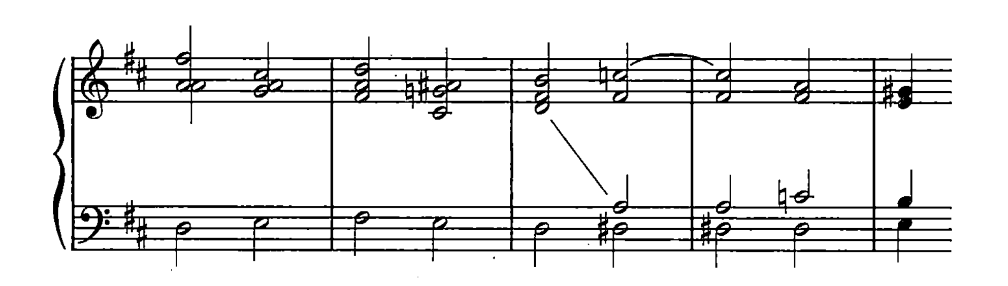

NAME ____________________

E. Bracket the deceptive resolution of a secondary dominant in this excerpt. What is unusual about the resolution chord here?

BEETHOVEN: SYMPHONY NO. 1 IN C MAJOR, II

F. After the initial tonic arpeggiation, the harmonies in measure 2 are somewhat surprising. What scale degree in measure 3 is tonicized? In the soprano voice leading, try connecting the G^5 in measure 1 to the C^5 in measure 4.

MENDELSSOHN: WEDDING MARCH FROM *A MIDSUMMER NIGHT'S DREAM*

NAME ______________________________

CHAPTER 26

Tonicization and Modulation III

MODULATION TO CLOSELY RELATED KEYS

1 The outer voices of a series of modulating progressions are given below. For each progression, determine the new closely related key area, which is reinforced by an authentic cadence, and provide a roman-numeral analysis of the harmonies leading to it. Be sure to indicate any pivot chords in terms of both keys. Do you notice any instances of chromatic modulation?

2 Realize the following figured-bass exercises, and supply each with a roman-numeral analysis. Exercise 2B is unfigured. Some passages modulate to but do not include an authentic cadence in the new key area. Several may employ a chromatic modulation, in which a pivot chord is not necessary.

E.

3 When harmonizing melodies of more than one phrase, it is important that you examine the different melodic formulas at the cadence for possible modulations. Each pair of soprano notes given below suggests more than one melodic cadence within the given key signature. For example, in Exercise 3A you might wish to set the B⁴–A⁴ in either A major or F♯ minor. Fill in the missing voices for the various two-chord cadences, and label the possible keys that are suggested with roman numerals.

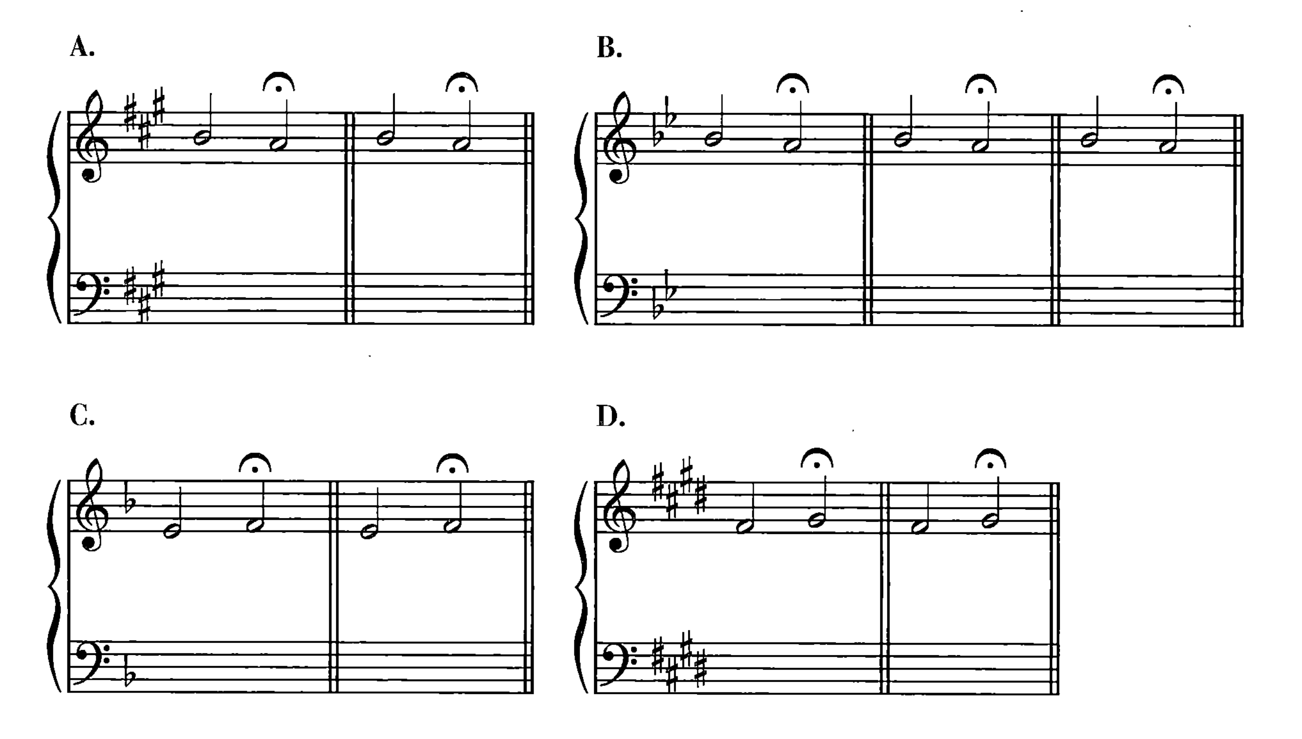

4 Make a four-voice setting of one of the tunes below. First, examine the cadence tones for possible tonicizations. Try to incorporate some modulations to closely related keys. For example, the first cadence in Exercise 4A might be an imperfect authentic cadence in F major ($\hat{2}$–$\hat{3}$ in the soprano) or a Phrygian cadence in D minor ($\hat{4}$–$\hat{5}$ in the soprano). Remember that the melody can return to the original tonic immediately following each cadence. Supply a roman-numeral analysis.

5 Analyze the following excerpts, all of which modulate to a new key area. In addition to supplying a roman-numeral analysis, provide a voice-leading reduction for the examples that are followed by blank staves.

A. Does the progression in measures 4–5 constitute a modulation? Be able to support your opinion.

ALEXEI LVOV: "GOD SAVE THE CZAR" (RUSSIAN NATIONAL ANTHEM)

B. Although the last chord omits the chordal 3rd, why is there no doubt about the mode of the new key?

MOZART: WIND SERENADE IN C MINOR, K.388, IV

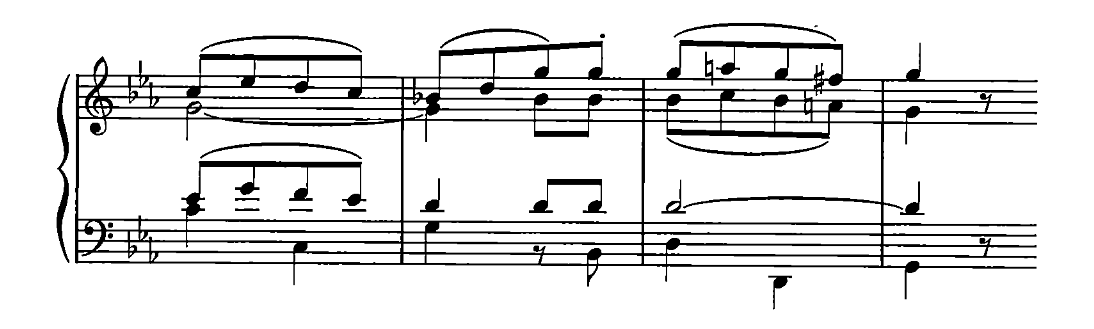

C. What type of sequence is employed in measures 1–6?

HANDEL: MINUET IN F MAJOR

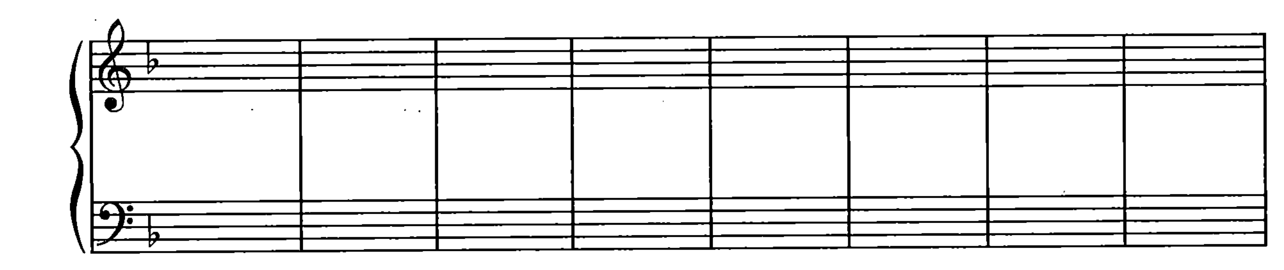

D. This quotation contains an extended tonicization of the minor dominant key (v), or E minor in terms of the original A major. Does a new secondary key area occur within this E-minor section? If so, what is its relation to E minor?

BEETHOVEN: STRING QUARTET IN A MAJOR, OP. 18, NO. 5, I

NAME ______________________

CHAPTER 27

Modal Exchange and Mixture Chords

1 Soprano lines have been provided for a series of short phrases. Write in the indicated mixture chord. Then complete the remainder of the progression in four-voice texture. Be sure to include a roman-numeral analysis.

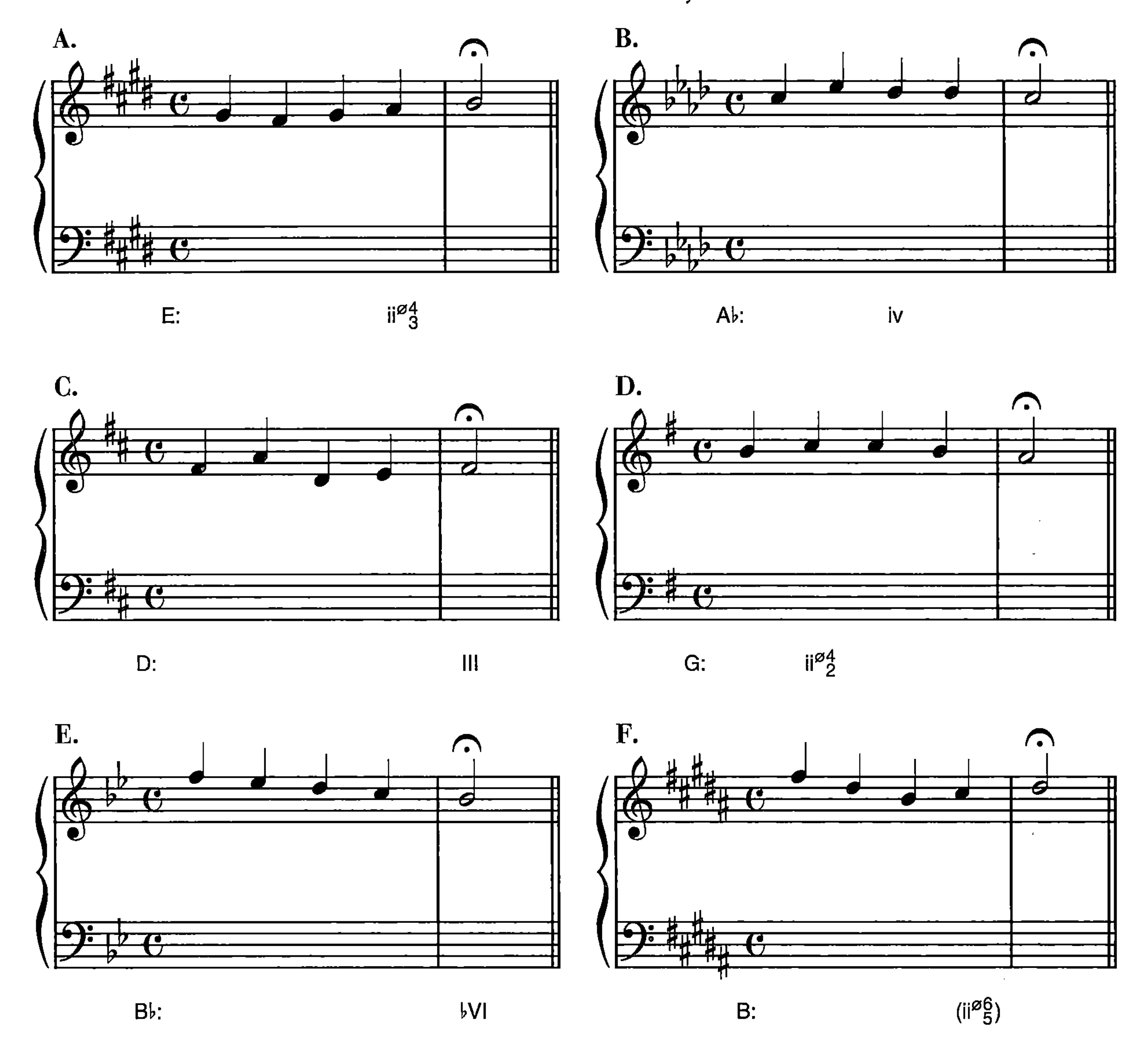

2 A short piece in chorale style is given below. After supplying a roman-numeral analysis, locate those harmonies that are likely mixture candidates; not all will work equally well. Change their notation and roman numerals to the desired mixture chords. Always consider the possibility of modulation. Play the original and your version with mixture chords, and compare the difference in effect.

3 Realize the following figured-bass exercises and give a roman-numeral analysis for each.

NAME ______________________________

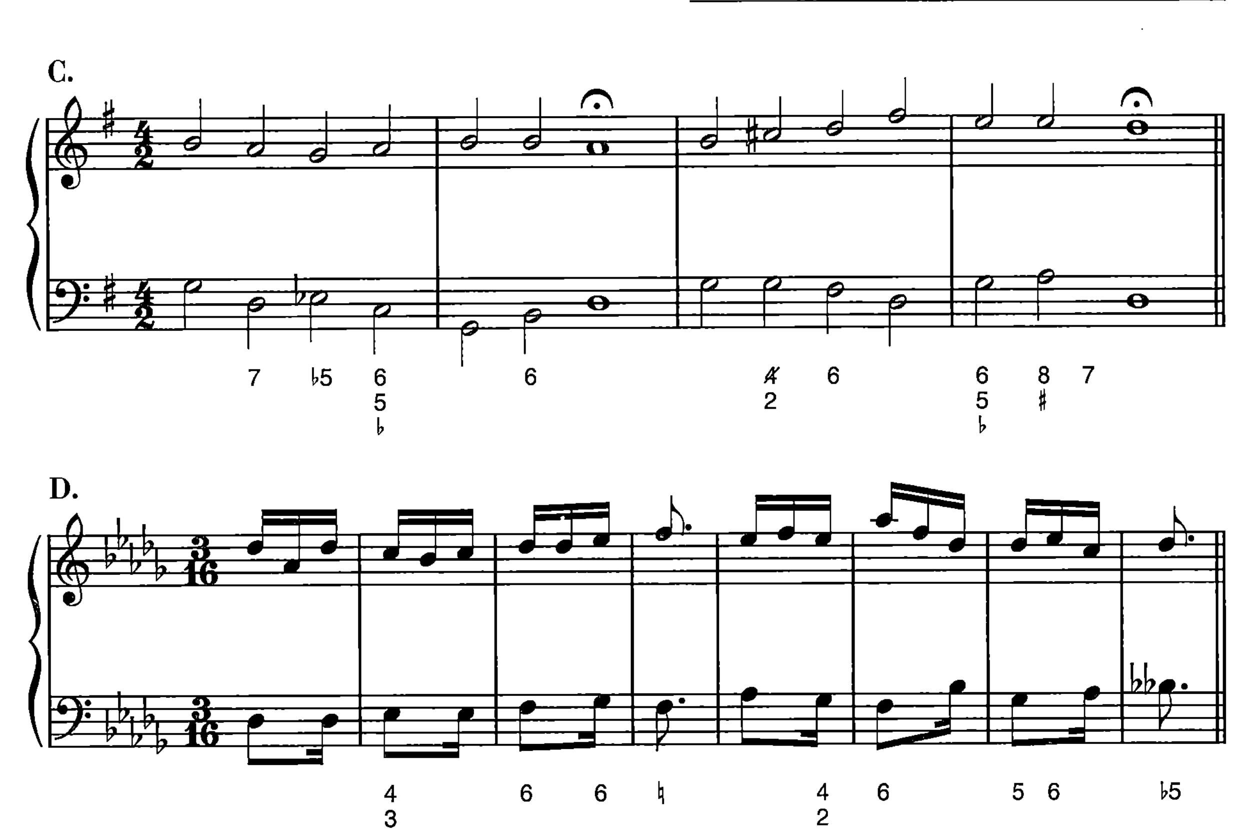

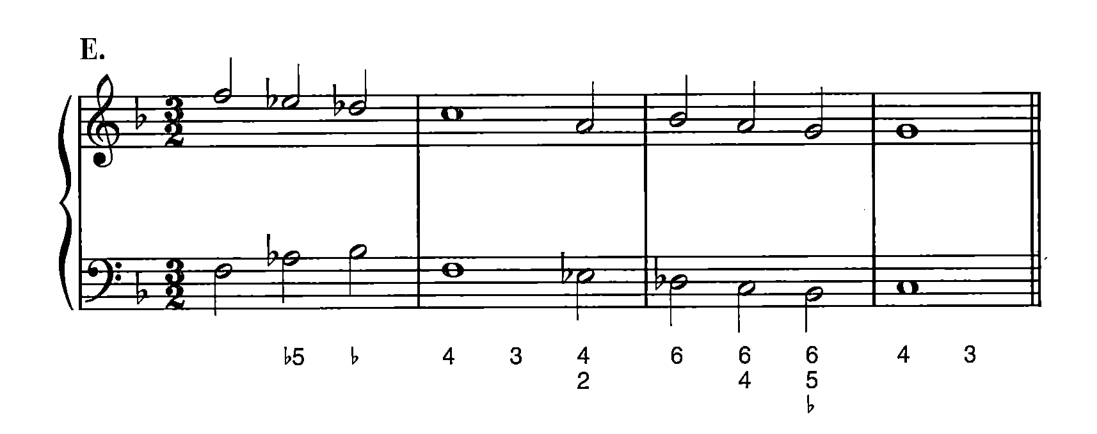

4 Give a roman-numeral analysis for each of the following excerpts, and provide a voice-leading graph for those that are followed by empty staves.

A. Describe the treatment of the nonharmonic tones in the upper voice.

DVOŘÁK: CELLO CONCERTO IN B MINOR, III

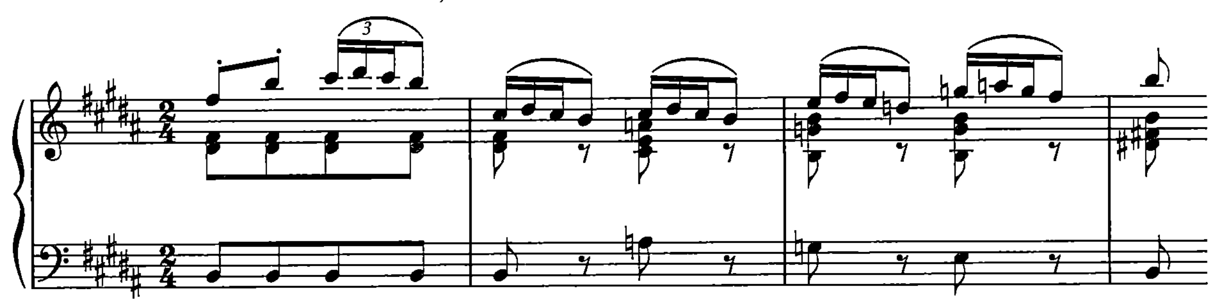

B. In what key do the mixture chords occur?

"National Hymn" (hymn)

C. Which mixture chord is prolonged in this passage?

Gounod: Trio from *Faust*, Act IV (simplified)

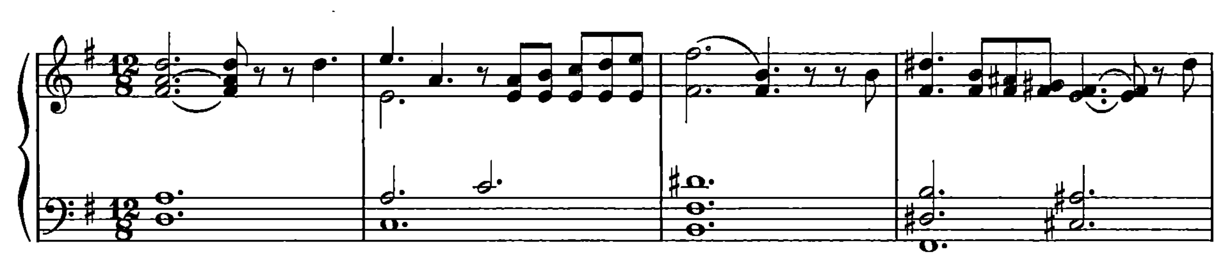

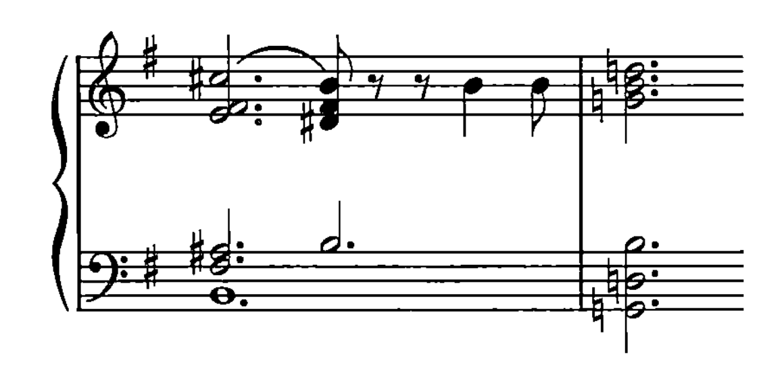

D. What key is momentarily tonicized in measures 3–4?

SCHUBERT: WALTZ IN F MAJOR

E. This kind of closing section is sometimes called an extended plagal cadence. Which mixture chord extends the plagal motion to I?

BRAHMS: SYMPHONY NO. 3 IN F MAJOR, II

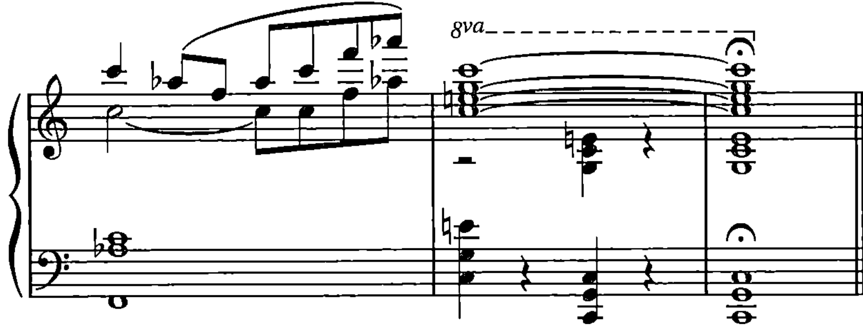

F. Although this beautiful passage contains a number of applied chords, what is unusual about its only mixture chord?

VERDI: LOVE DUET FROM *OTELLO*, ACT I

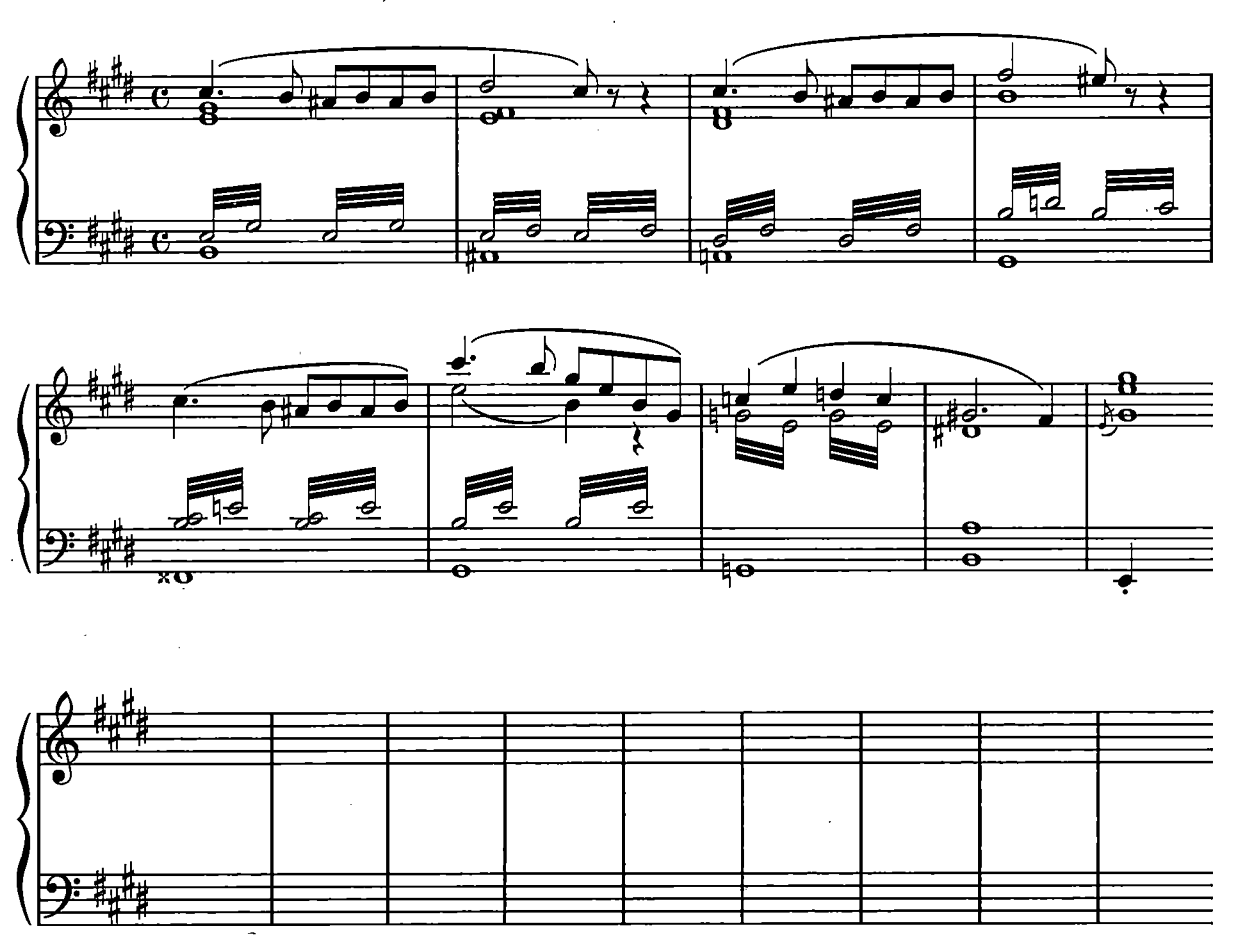

G. This sweeping passage provides an impetuous nontonic opening to Strauss's portrait of the legendary libertine. What is the first chord?

RICHARD STRAUSS: *DON JUAN*

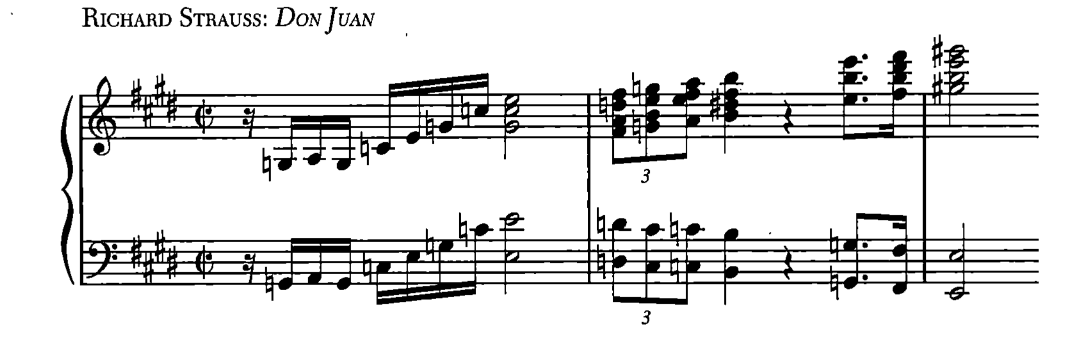

NAME ______________________________

CHAPTER 28

The Neapolitan Chord

1 Write out cadences for each of the following passages. Either the soprano or bass voice is provided. Employ an appropriate Neapolitan chord in each. For Exercises 1B and 1C, incorporate the chords indicated by the symbols in your settings.

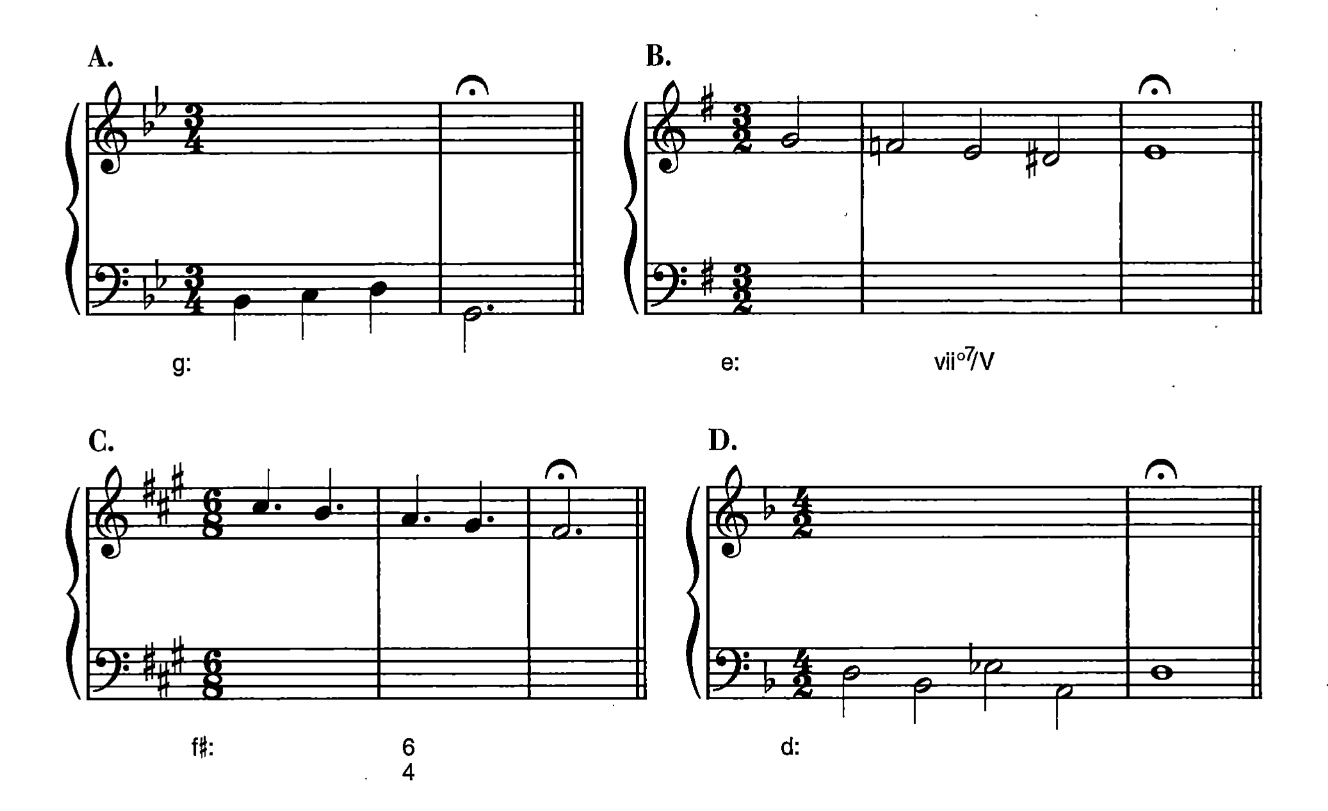

2 Realize the following figured-bass exercises, and supply a roman-numeral analysis for each. Exercise 2A is unfigured. Although the Neapolitan chords in most of the exercises are cadential, several are found in embellishing progressions within the phrase. How can you avoid the potential parallels in Exercise 2D?

NAME ______________________________

3 Harmonize this melody, using appropriate Neapolitan chords.

4 Give a roman-numeral analysis of the short excerpts cited below. You may wish to make a voice-leading reduction of some passages on a separate sheet.

A. How is the Neapolitan sixth treated here?

BRUCKNER: *CHRISTUS FACTUS EST* (TEXT OMITTED)

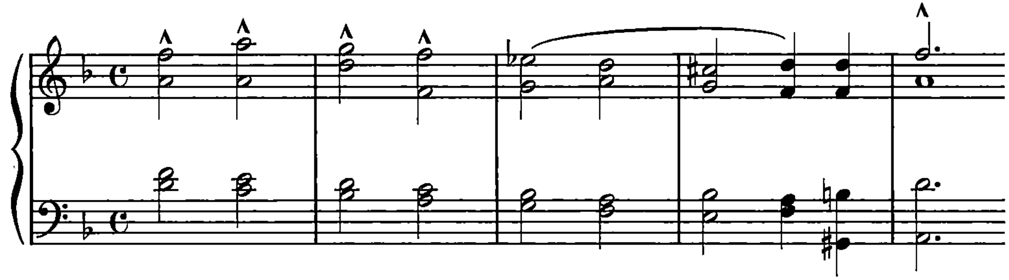

B. What is unusual about the resolution of the chord on the second beat of measure 2?

MOZART: STRING QUARTET IN D MINOR, K.421, IV

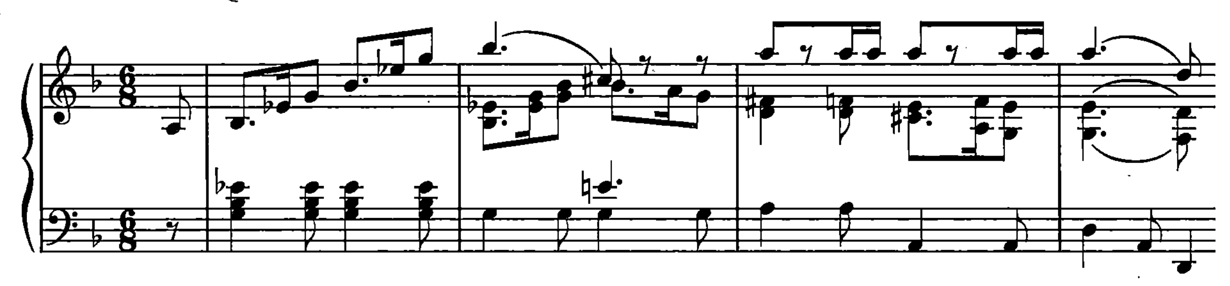

C. What creates a dissonant clash with the Neapolitan chord (in the key of B minor) in measure 4? What interval is emphasized? Why do you think this interval is musically appropriate in this instance? (The title of the excerpt gives a clue.)

WAGNER: THE CURSE ON THE RING FROM *DAS RHEINGOLD*, SCENE 4

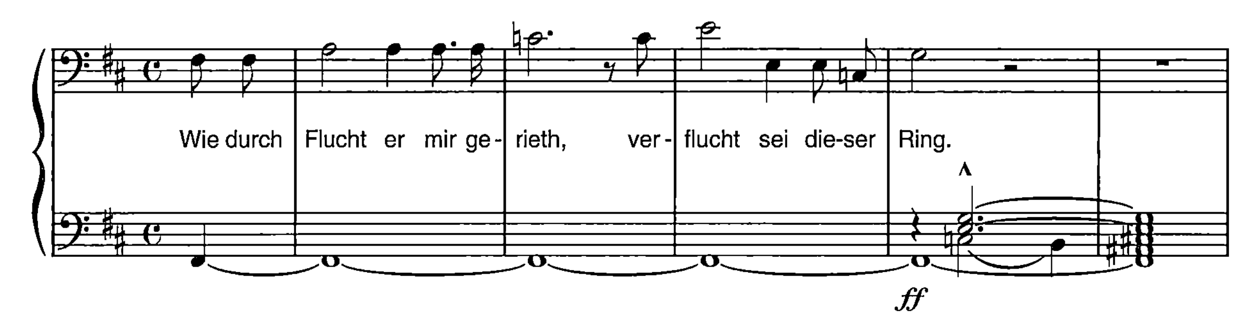

Wie durch Flucht er mir gerieth,
verflucht sei dieser Ring.

Since through a curse I gained it,
My curse lies on this ring.

D. Where is the ♭II and how is it tonicized?

CHOPIN: MAZURKA IN A MINOR, OP. 7, NO. 2

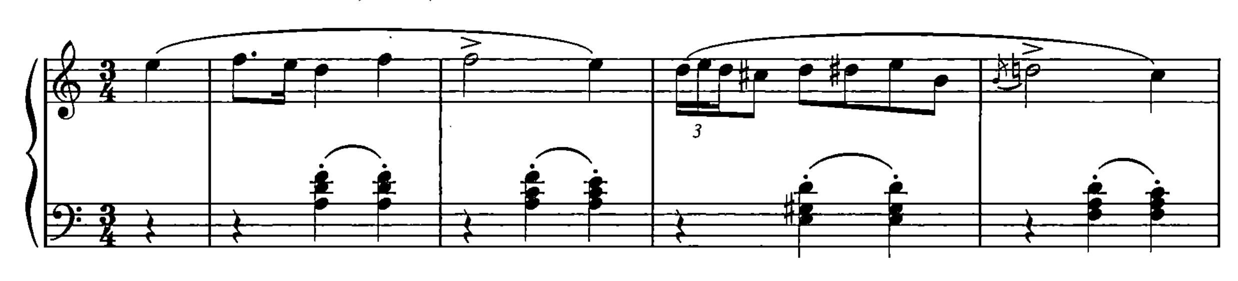

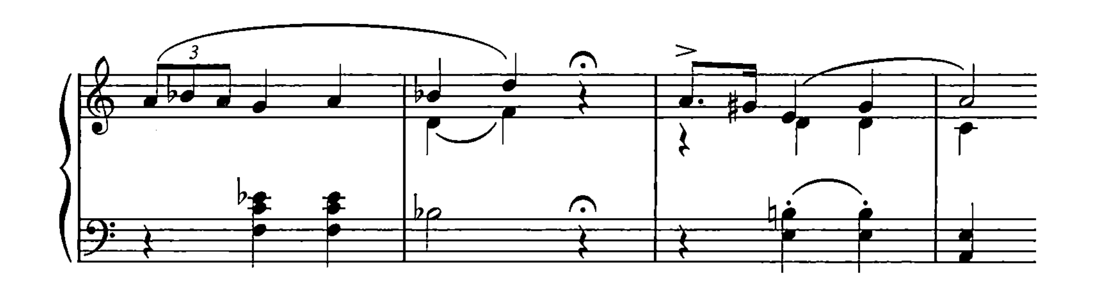

E. Why do you think the composer changed the key signature in this passage?

SCHUBERT: MOMENT MUSICAL NO. 6 IN A♭ MAJOR

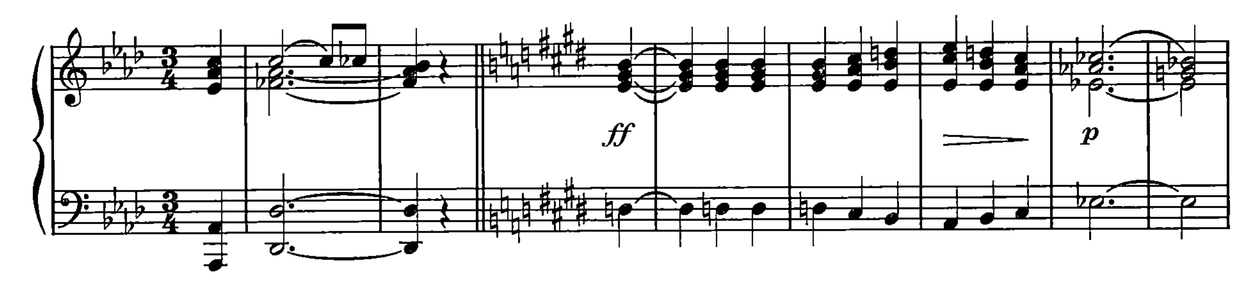

F. What is peculiar about the Neapolitan chord in this excerpt?

SCHUBERT: STRING QUARTET NO. 14 IN D MINOR ("DEATH AND THE MAIDEN"), I

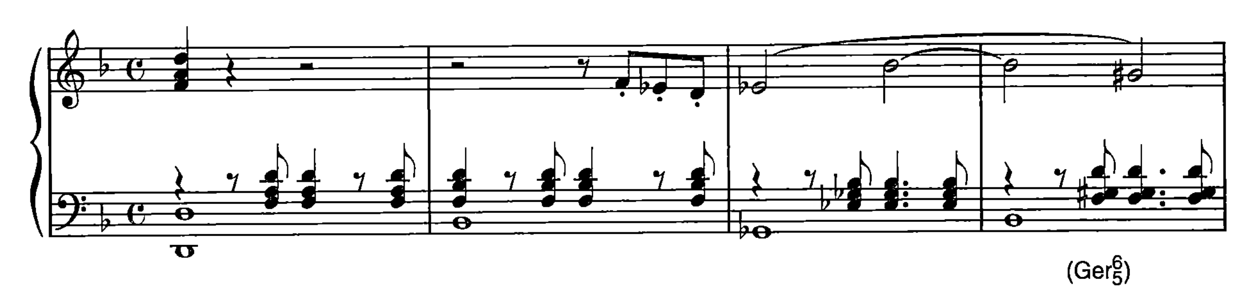

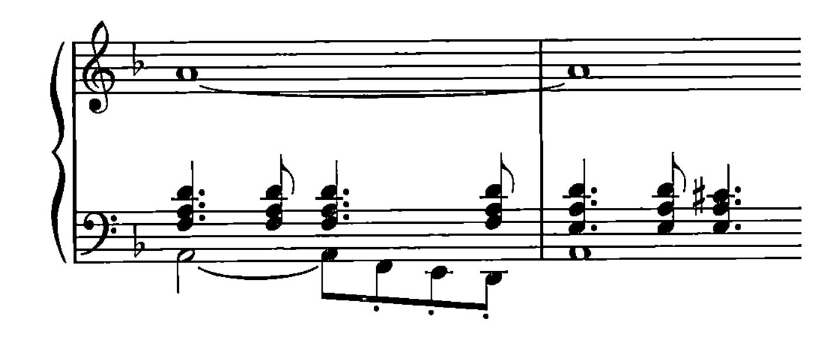

G. What gives an additional "bite" to the ♭II harmony in measures 4–5?

SAINT-SAËNS: *INTRODUCTION AND RONDO CAPRICCIOSO*

H. In this curious passage, why does the harmonic background in measure 2 seem at odds with the soprano line? What two different harmonic progressions are suggested?

BEETHOVEN: PIANO SONATA IN C♯ MINOR ("MOONLIGHT"), OP. 27, NO. 2, I

NAME ______________________

CHAPTER 29

Augmented Sixth Chords

1 In four-voice texture, approach and resolve the following Aug6th chords, using the roman-numeral labels that are supplied; watch for different inversions. The soprano voice is given, but the bass part must be deduced from the roman numerals and figured-bass symbols. Explain the derivation of the Aug6th in each progression.

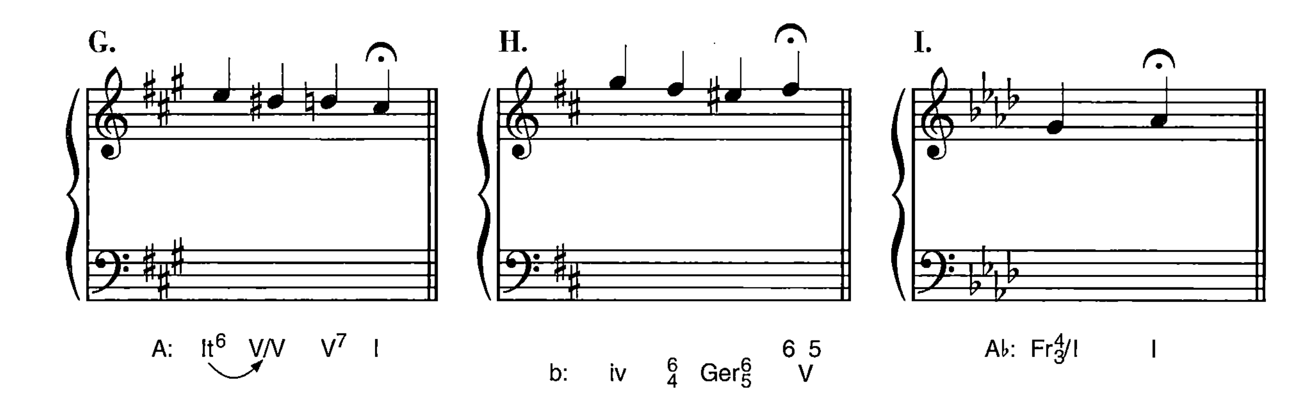

2 Realize the figured-bass exercises and supply a roman-numeral analysis of each. In Exercise 2C, only the beginning chord and figures for the first Aug6th chord are given. Be careful how you label and resolve the various Aug6th chords.

D.

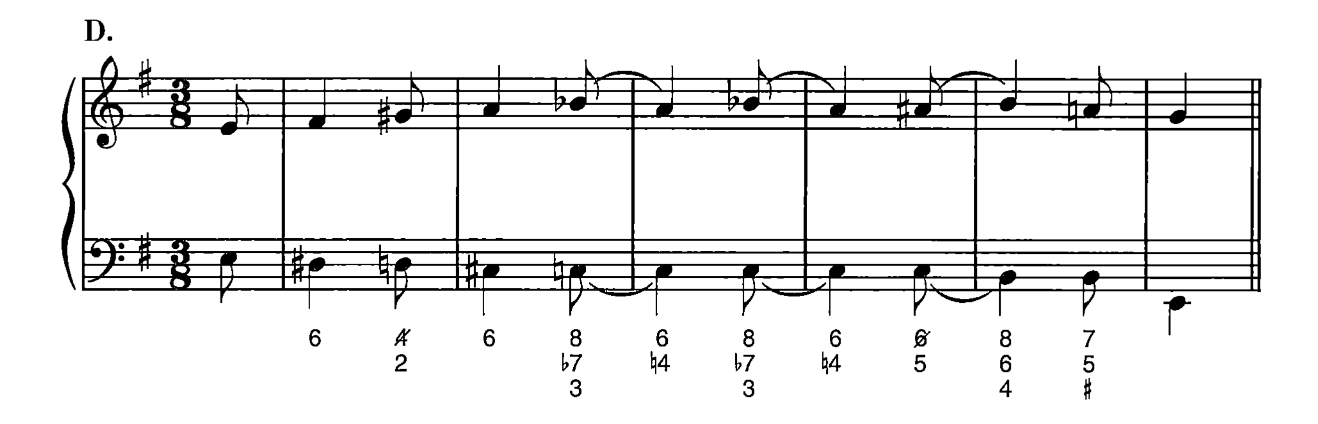

3 Supply a roman-numeral analysis of these excerpts and provide voice-leading reductions for those passages followed by empty staves.

A. Locate, identify, and explain the derivation of the Aug6th in this "enigmatic" theme.

ELGAR: THEME FROM *ENIGMA VARIATIONS*

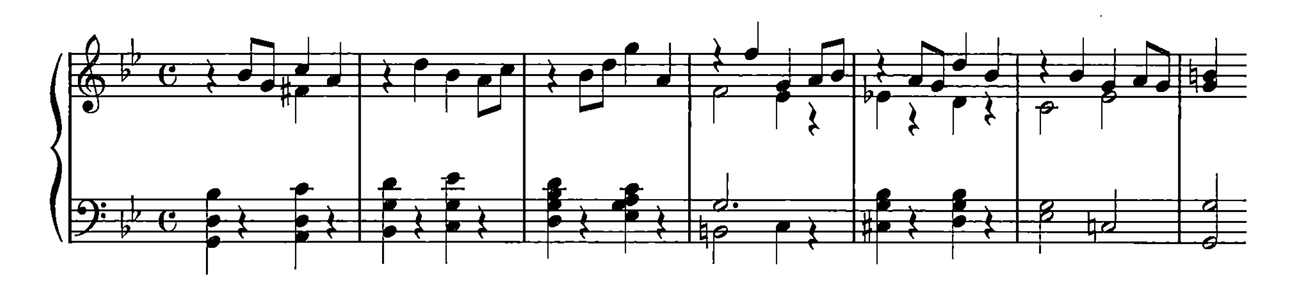

B. What voice-leading device is employed in this brief passage? How would you indicate it in your reduction?

MOZART: DUEL SCENE FROM *DON GIOVANNI*, ACT I

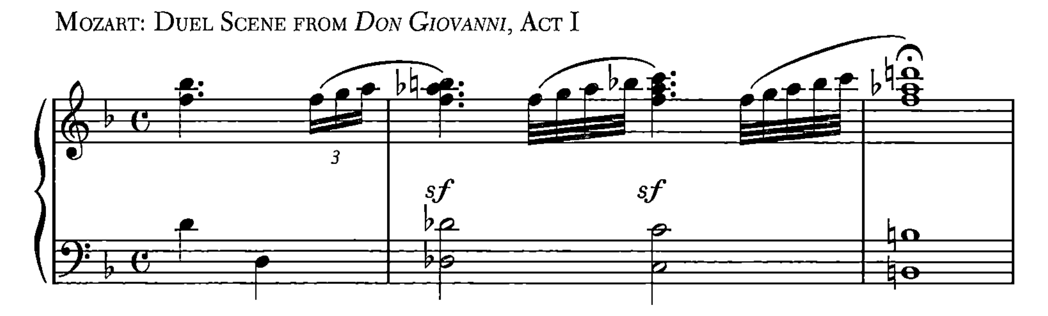

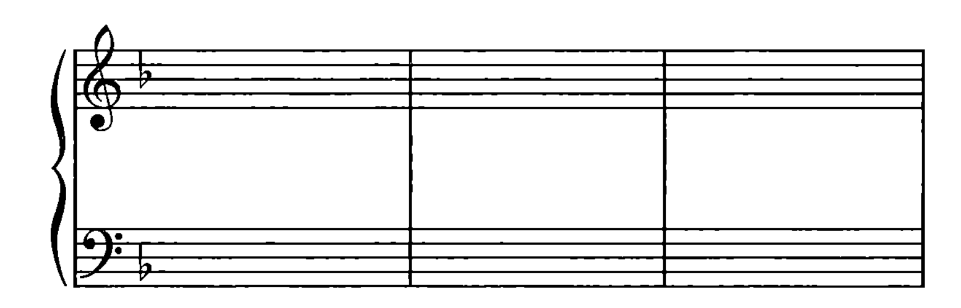

C. What type of chord precedes each basic harmony on the third beat of measures 1–3? How is the final dominant approached?

SCHUBERT: STRING QUINTET IN C MAJOR, I

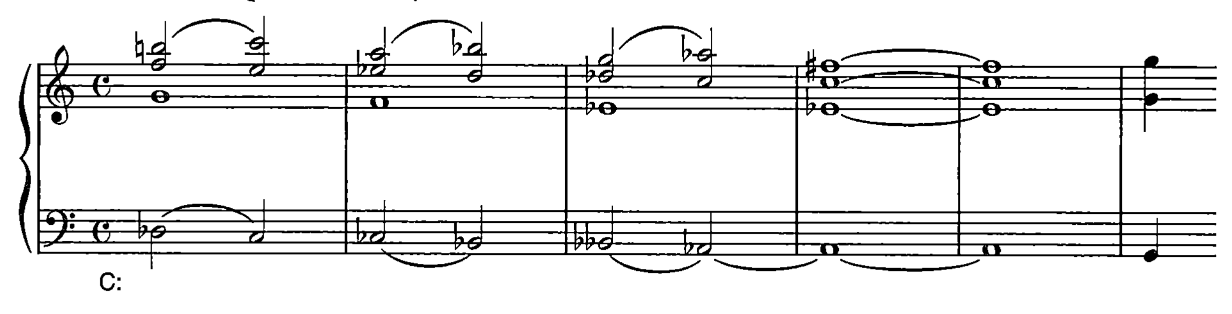

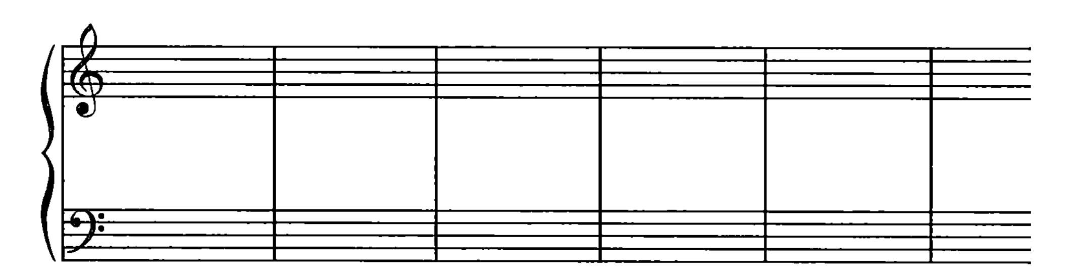

D. How is the listener harmonically "misled" at the opening of this charming waltz?

SCHUMANN: "WALTZ" FROM *ALBUMBLÄTTER*

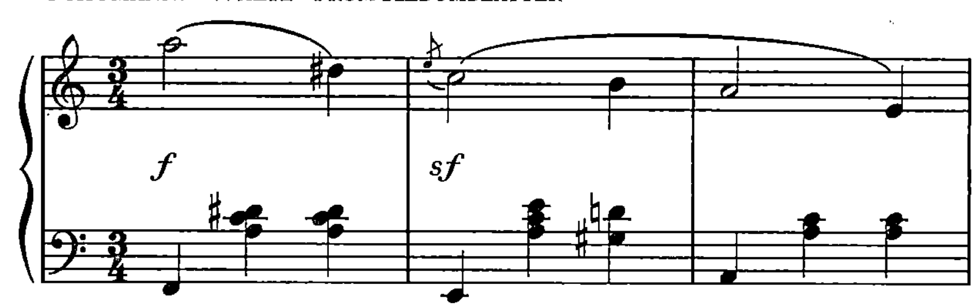

E. In the reduction that is given, what two chords immediately precede the final tonic? What do we call this relationship?

WAGNER: OVERTURE TO *TANNHÄUSER*

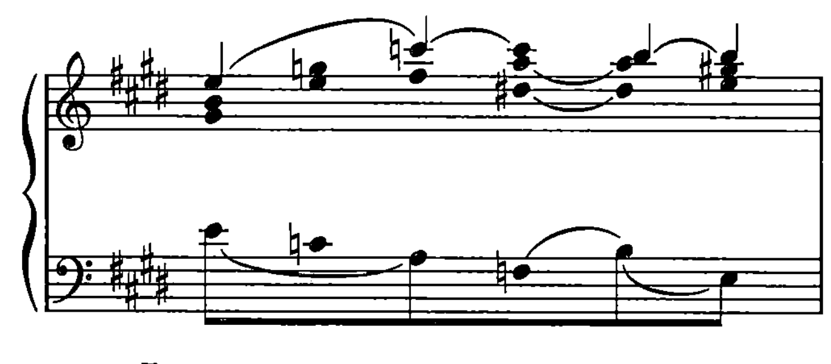

F. Specify why the resolution of the Aug6th chord is so odd.

BRAHMS: INTERMEZZO IN A MINOR, OP. 118, NO. 1

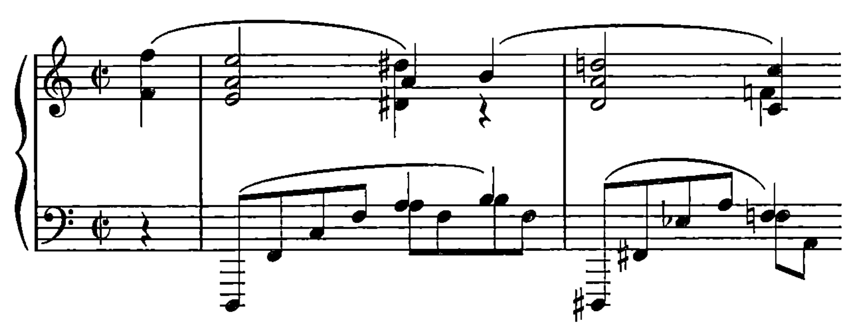

a:

G. The use of the "Tristan chord" as an Aug6th is discussed in the text. In this excerpt, another harmony is appended to Wagner's basic two-chord progression. How does the chord in measures 2–3 now function in light of this appended chord? (You may wish to spell it enharmonically.)

WAGNER: KING MARK'S MONOLOGUE FROM *TRISTAN UND ISOLDE*, ACT II

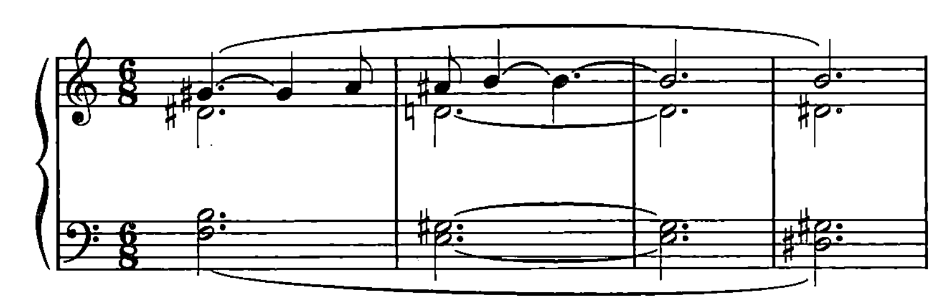

4 An extensive passage from Schumann's "Die beiden Grenadiere" (The Two Grenadiers) is quoted below. Make a voice-leading reduction, pointing out the use of Aug6th chords. There are actually *two* different ways of analyzing this section; you may wish to discuss the merits of each in class.

SCHUMANN: "DIE BEIDEN GRENADIERE," OP. 49, NO. 1

Wun - de." Der An - dre sprach: "Das Lied ist aus, auch ich möcht' mit dir
ster - ben, doch hab' ich Weib und Kind zu Haus, die oh - ne mich ver - der - ben."

E X C U R S I O N 2

More Complex Forms

As in Excursion I, a number of pieces or movements for possible analysis are listed below.

Sonata Form

Clementi: Sonata in D major, Op. 36, No. 6
Haydn: Symphony No. 98 in C major, I
Haydn: Symphony No. 100 in G major ("Military"), I. After the lengthy introduction, this movement is an example of a monothematic sonata form.
Mozart: Piano Sonata in C major, K.545. What is unusual about the recapitulation of this movement?
Beethoven: Piano Sonata in F minor, Op. 2, No. 1
Beethoven: Symphony No. 5 in C minor, I. This movement ends with an extended coda.

The opening movements of the Beethoven Piano Sonatas in D minor, Op. 31, No. 2 ("Tempest"), and C major, Op. 53 ("Waldstein"), contain some interesting deviations from conventional sonata form. Either might form the basis for an extended analytical paper.

Concerto Form

Mozart: Concerto for Flute in G major, K.285c
Beethoven: Concerto for Piano No. 3 in C minor, I

Sonata-Rondo Form

Mozart: Piano Sonata in B♭, K.333, III
Mendelssohn: Scherzo from *A Midsummer Night's Dream*

Chorale Prelude

Bach: "Wachet auf" (Schübler Chorales for organ)
Bach: "Jesu, Joy of Man's Desiring"
Bach: "Vor deinen Thron tret' ich"

Invention

Bach: Two-Part Invention in F major
Bach: Two-Part Invention in B minor
Bach: Three-Part Invention in D major

Fugue

Handel: "And with His Stripes We Are Healed" from *Messiah*
Bach: Fugue in C minor from *Well-Tempered Clavier,* Book I
Bach: Fugue in G minor from *WTC* I
Bach: Fugue in D minor from *WTC* II
Mozart: Kyrie from *Requiem*

NOTES

NAME ______________________

CHAPTER 30

Implication and Realization

1 A short composition from Schumann's collection of piano pieces for children is quoted below. As you play or listen to it on CD3, trace how the composer continually uses standard progressions to set up harmonic/melodic expectations, only to substitute unexpected resolutions or continuations. Try playing the chords that you anticipate, and then playing Schumann's actual version.

In addition to providing a roman-numeral analysis, you might wish to make a voice-leading reduction in the empty staves that are given. Notice the opening $\hat{3}$–$\hat{2}$–$\hat{1}$ line in the melody. The final phrase is internally extended by two measures. What is the relationship of this extension to the cadential $^6_4$ that begins in measure 15? How is this $^6_4$ related to the one in measure 17, which ushers in the $\hat{3}$–$\hat{2}$–$\hat{1}$ resolution in the soprano?

SCHUMANN: "***" FROM *ALBUM FOR THE YOUNG*, OP. 68, NO. 21

9
a tempo
13
p
Etwas langsamer

NAME ______________________________

CHAPTER 31

Ninth, Eleventh, Thirteenth, and Added-Note Chords

1 On the following staves, the key, some type of ninth or eleventh chord, and a soprano note are supplied. Make a five-voice setting of a three-chord progression for each; beware of parallels! Approach and resolve the chordal 9th or 11th stepwise, using either neighboring (N) or suspension (S) figures as indicated.

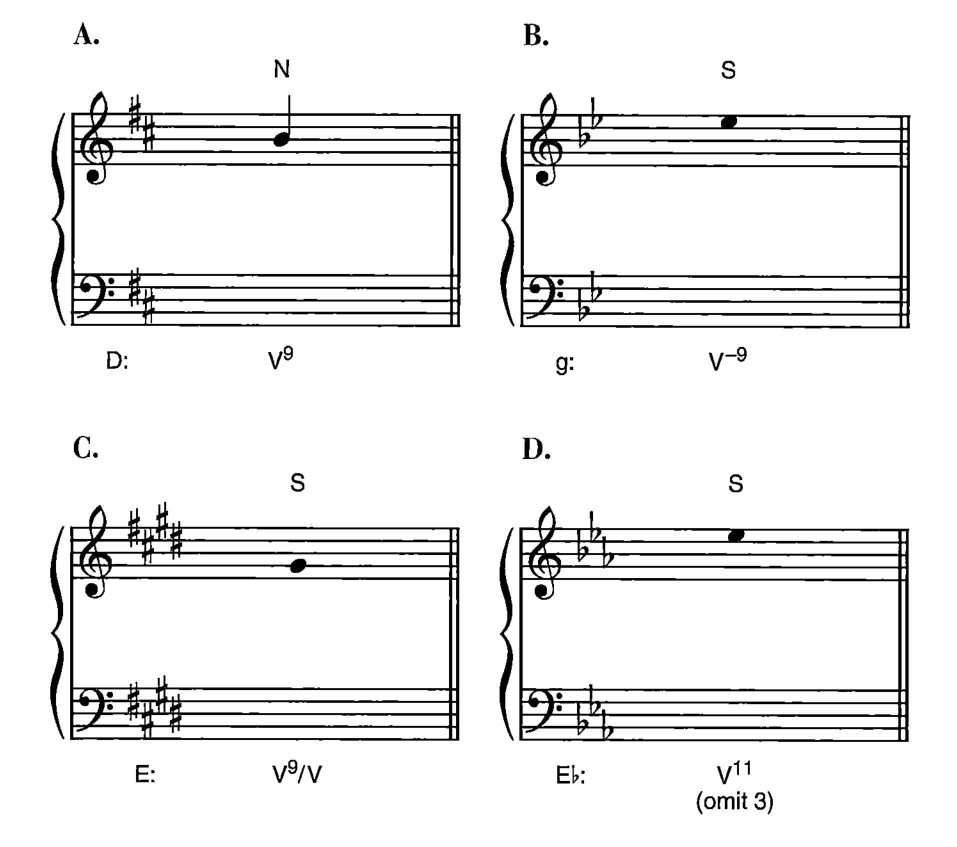

2 Realize the figured-bass exercises below in *four-voice* texture. This will necessitate the omission of the chordal 5th in all ninth chords. Supply a roman-numeral analysis for all passages except the last.

A.

B.

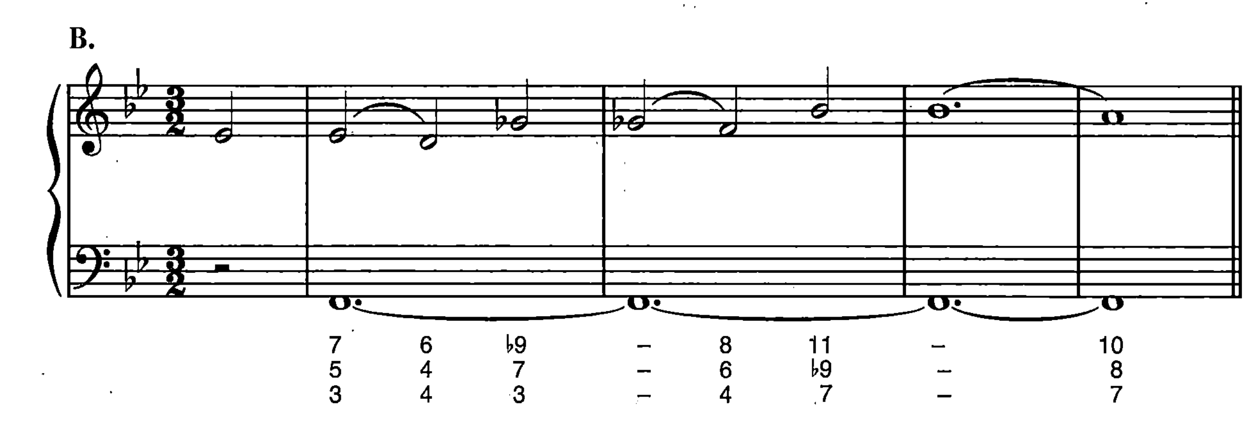

C.

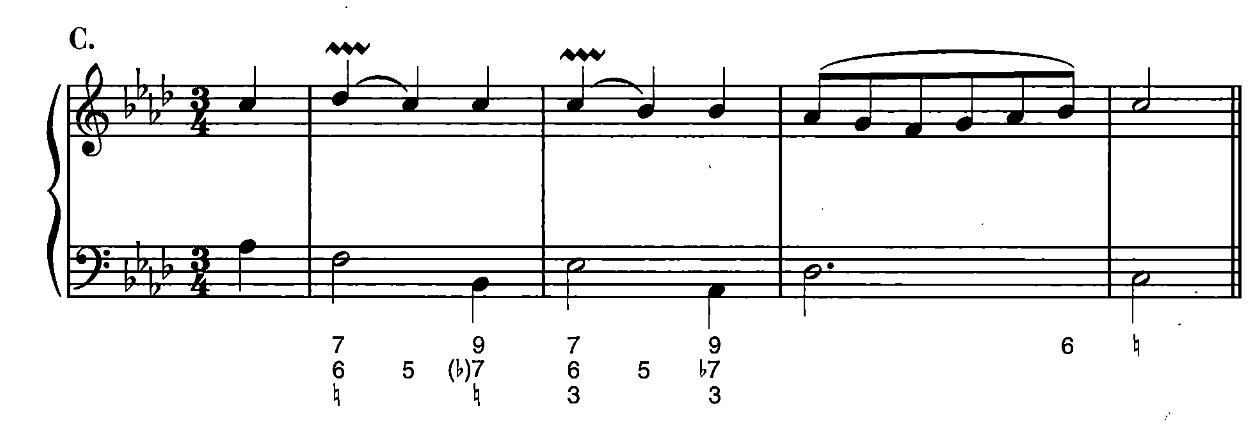

3 A harmonic progression has been indicated with commercial chord symbols (you might wish to consult Appendix 4 of the text). Write out the bass line on the empty measures and provide roman numerals.

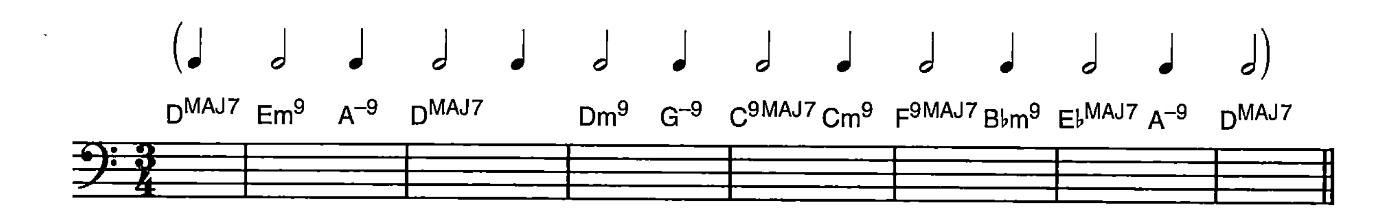

4 Give a roman-numeral analysis of the following excerpts, providing a voice-leading reduction for those passages supplied with empty measures.

A. What melodic figuration does the approach to and resolution of the chordal 9th in measures 6–7 resemble? (Consider the possibility of octave displacement.)

JOHANN STRAUSS, JR.: *WINE, WOMEN, AND SONG* (WALTZ)

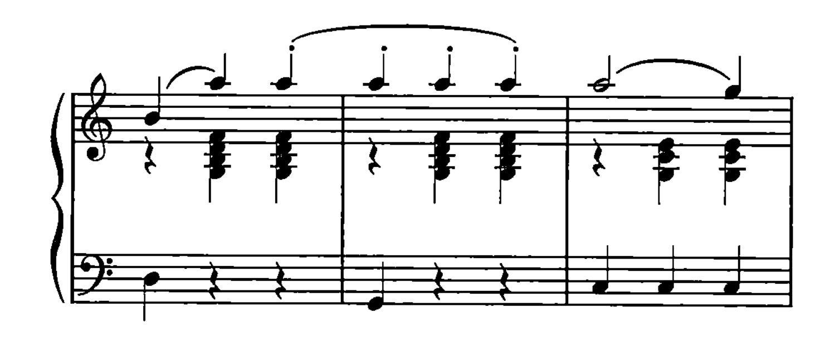

B. Notice that both the V^9 and V^{-9} are employed in this dominant prolongation. After making your reduction of this passage, state how the chordal 9ths are approached.

BEETHOVEN: PIANO SONATA IN B♭ MAJOR, OP. 22, III

C. What is the function and chord type of the ninth chord in this passage?

SCHUMANN: "SCHEHERAZADE," FROM *ALBUM FOR THE YOUNG*, OP. 68, NO. 32

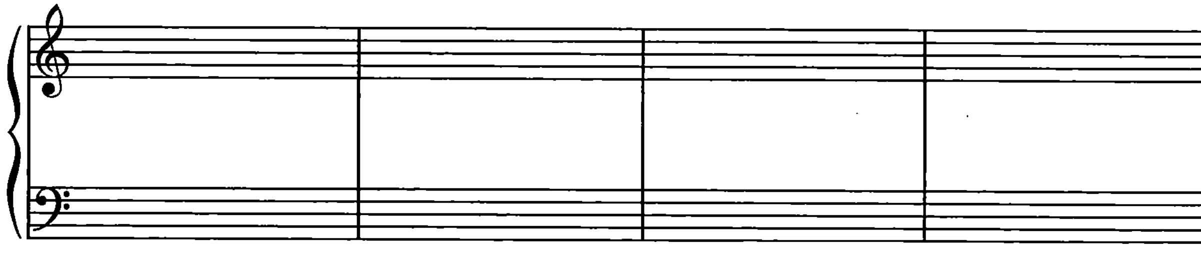

D. The upper voice may be thought of as a compound melody made up of two separate lines. How can you explain the curious dissonance in the fourth measure?

BEETHOVEN: ECOISSAISE IN G MAJOR, WoO 23

E. The opening of this song contains a variety of ninth chords. What is the familiar harmonic sequence that underlies this progression?

Fauré: "Après un rêve," Op. 7, No. 1

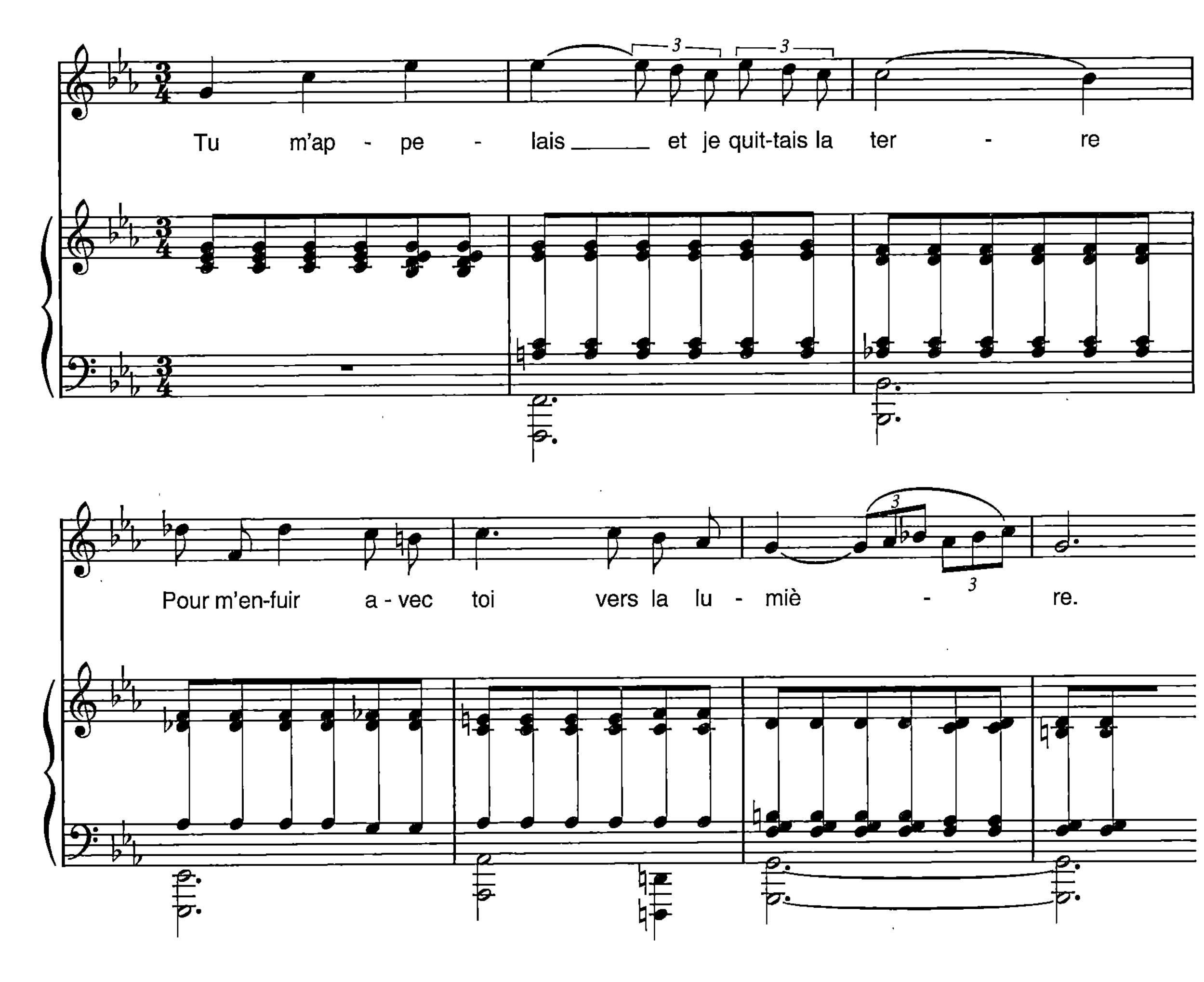

NAME ______________________________

CHAPTER 32

Embellishing Chromatic Chords

1 **A.** Using the empty staves below, employ the diminished seventh chord D♯ F♯ A C in C major in each of the following functions, using a minimum of three chords:

1. a secondary vii^{o7}
2. an embellishing neighboring chord
3. an embellishing passing chord

You may need to employ enharmonic notation in some cases.

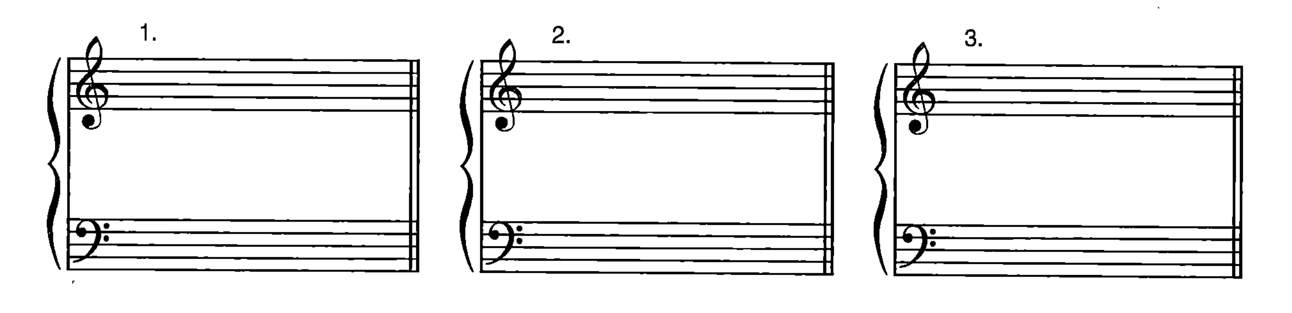

B. Now employ the German 6th F A C D♯ in A major in each of the following functions:

1. a normal Ger^6_5
2. an embellishing neighboring chord
3. an embellishing passing chord

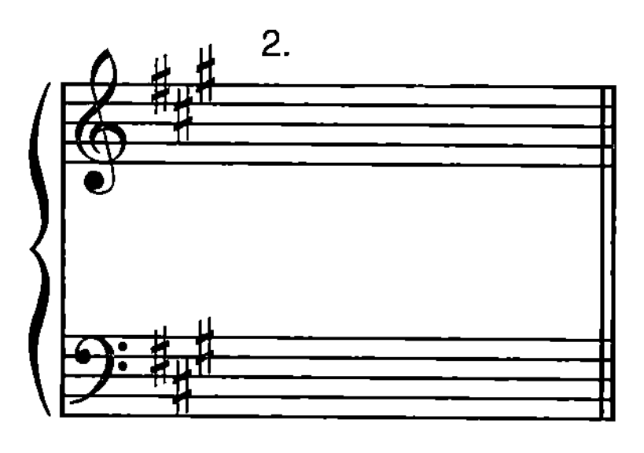

2 Write out two authentic cadences in keys of your own choice, using four-voice texture. Employ a V^{o7} and V^{+7}, respectively. Be careful about the partwriting of the augmented 6ths and diminished 3rds and their resolutions.

3 Realize the following figured-bass exercises and supply a roman-numeral analysis of each. Be careful of the partwriting in Exercise 3A. Exercise 3D contains a variety of less common chromatic chords.

C.

D.

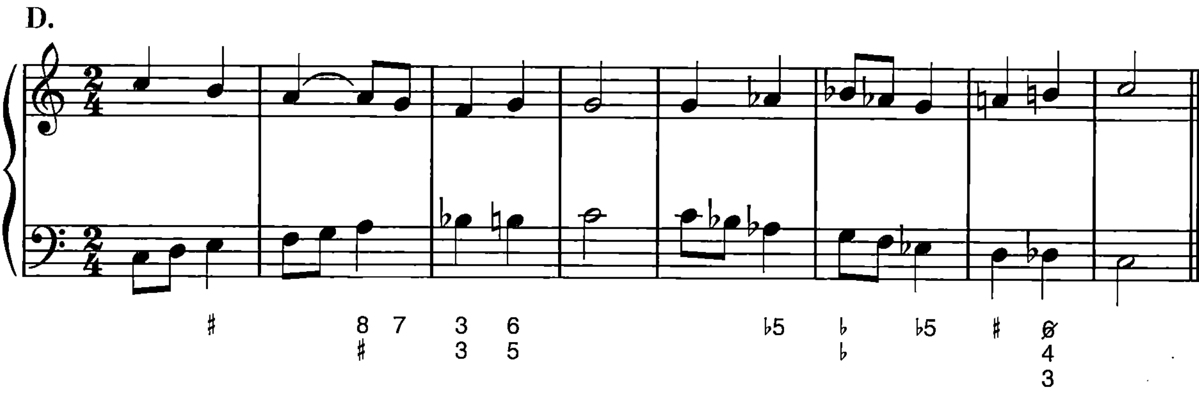

4 Provide a roman-numeral analysis for the following excerpts.

A. Make a voice-leading reduction of this passage in the empty staves that are provided. Explain the harmony in measure 2.

DVOŘÁK: SYMPHONY NO. 9 IN E MINOR ("NEW WORLD"), II

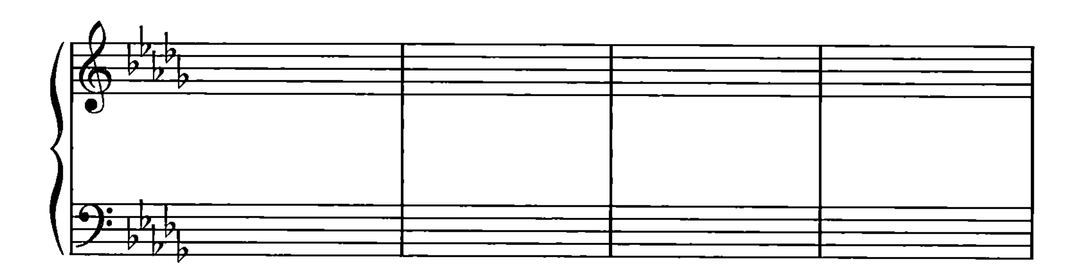

B. Observe how the embellishing chords allow a chromatic ascent in the vocal line from $G\sharp^4$ to $C\sharp^5$. What is the missing chordal member on beat 2 of the second measure?

SCHUBERT: "TÄUSCHUNG" FROM *WINTERREISE*

C. How does the function of the embellishing chord in measure 2 differ from that in measure 5? Using the empty staves, make a reduction in *five voices* (four parts in the upper staff and one in the bass).

TCHAIKOVSKY: WALTZ FROM *SLEEPING BEAUTY*, ACT I

D. How do the embellishing harmonies function in this lovely aria?

BIZET: MICAELA'S ARIA FROM *CARMEN*, ACT III

E. What two different ways do these short passages from Rachmaninoff's famous Prelude treat a ♭$\hat{2}$ in the soprano voice?

RACHMANINOFF: PRELUDE IN C♯ MINOR, OP. 3, NO. 2

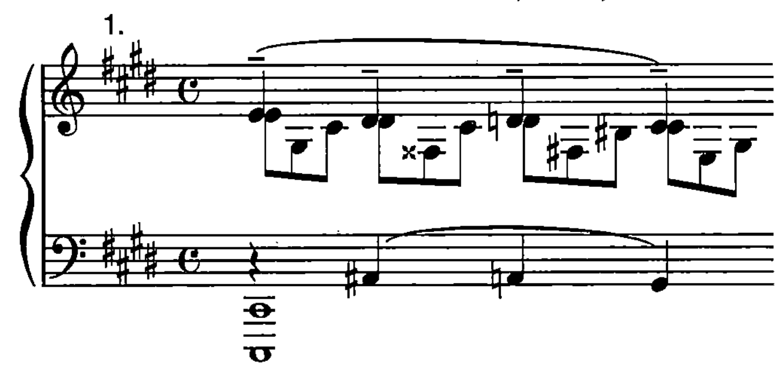

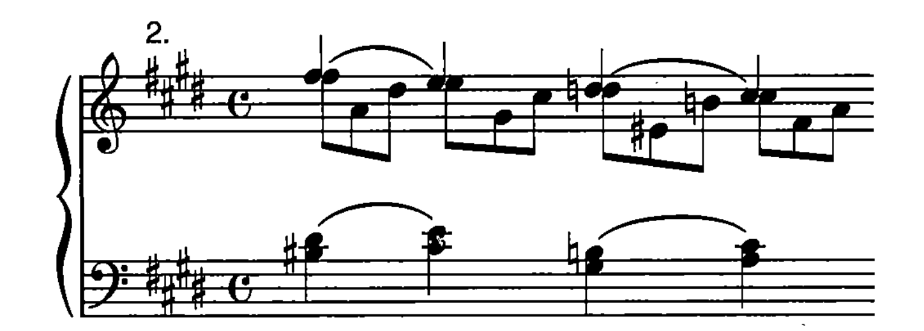

5 Compose a harmonization for one of the following phrases, utilizing at least three chromatic chords.

A.

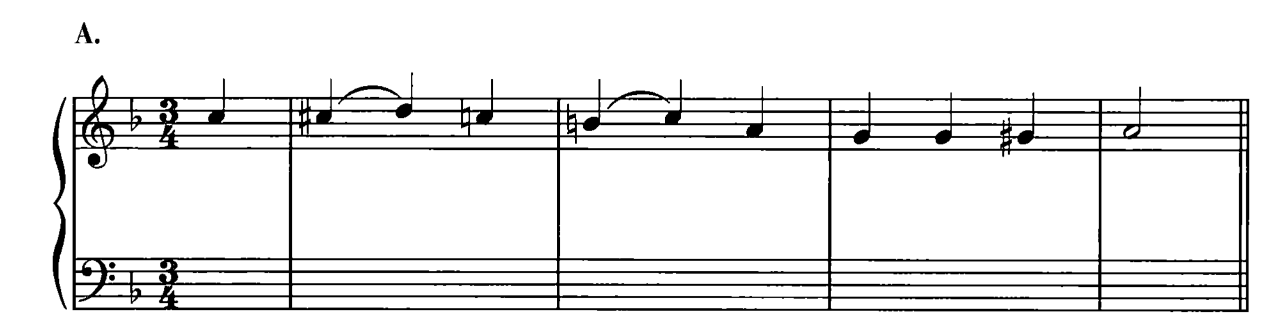

B.

NAME ____________________

CHAPTER 33

Dominant Prolongation

1 The following excerpts contain various dominant prolongations that occur in different formal contexts. Analyze each excerpt to determine the methods by which this harmony is extended. Supply voice-leading reductions in the empty staves that are provided.

A. The underlying V^7 in this passage is ornamented with chromatic embellishing harmonies, which need not form part of your analysis. Circle all ninth and eleventh chords.

WAGNER: RHINEDAUGHTERS' SCENE FROM *GÖTTERDÄMMERUNG*, ACT III

B. In this extensive retransition from a Haydn minuet, a variety of harmonies occurs over a dominant pedal. What happens in measures 8–13?

HAYDN: SYMPHONY NO. 100 IN G MAJOR ("MILITARY"), III

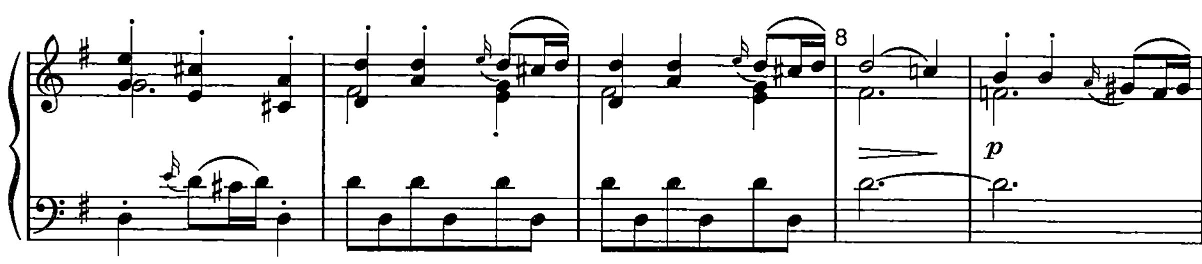

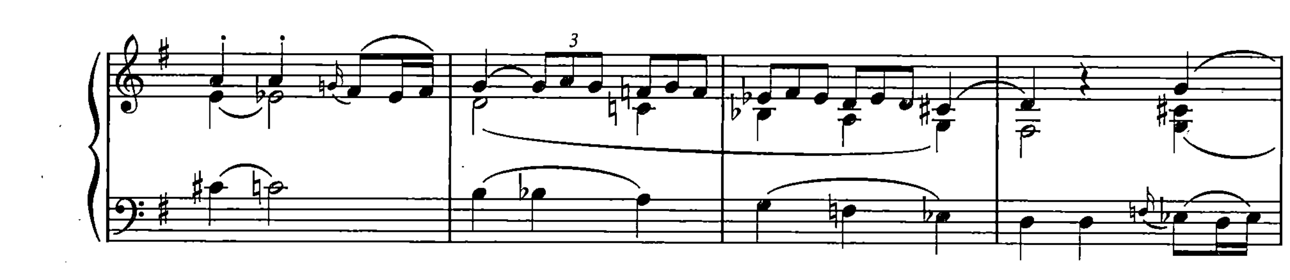

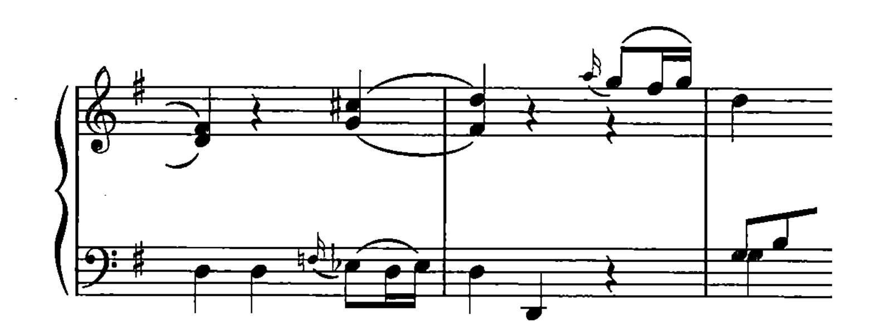

NAME ________________________________

C. Explain the different ways that these two passages prolong a cadential $^6_4$ before its eventual resolution to the tonic.

EDWARD MACDOWELL: "UNCLE REMUS" FROM *WOODLAND SKETCHES*

Verdi: "Miserere" from *Il Trovatore*, Act IV

NAME ________________________________

D. The opening sixteen measures of this waltz provide an introduction to the main theme. What does your voice-leading reduction reveal about how the dominant harmony is prolonged in measures 3–4 and 7–8?

CHOPIN: *GRANDE VALSE BRILLANTE* IN A♭ MAJOR, OP. 34

2 The return of the opening theme of a Minuet in Classical style is given below. In the blank measures, compose a retransition to this thematic material that contains a dominant prolongation. Try to retain the same style, and perhaps even utilize some of the motivic ideas.

CHAPTER 34

Modulation to Foreign Keys I

1 Realize the following figured-bass exercises and provide a roman-numeral analysis of each. Each passage employs at least one instance of a modulation to a foreign key. In the space provided, name the type of modulation used.

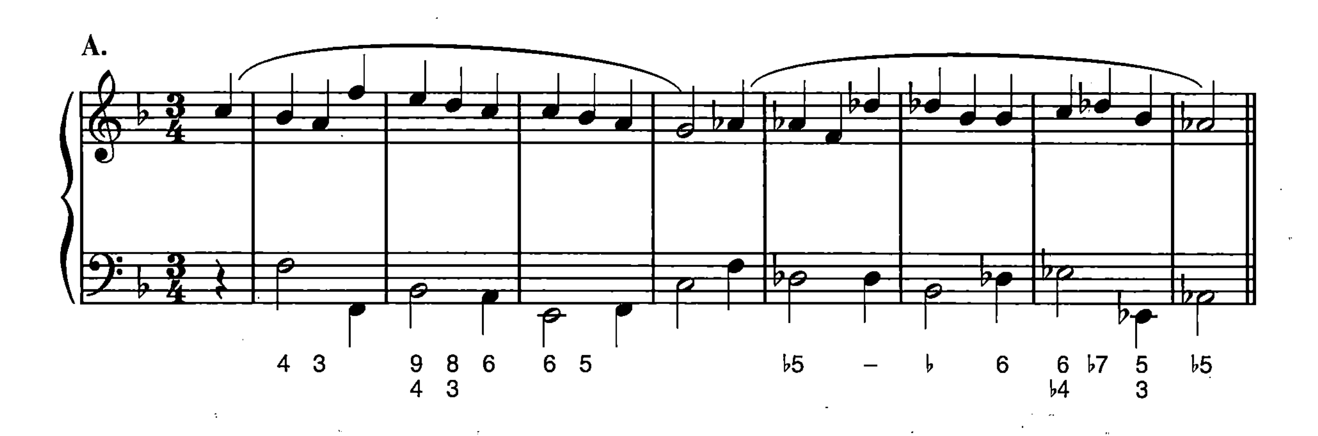

____

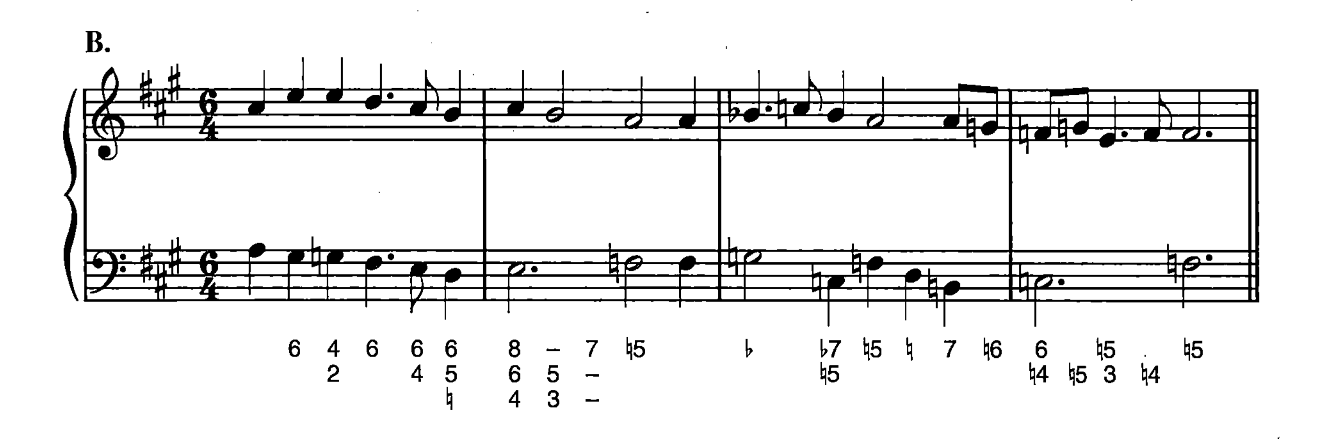

____

C.

D.

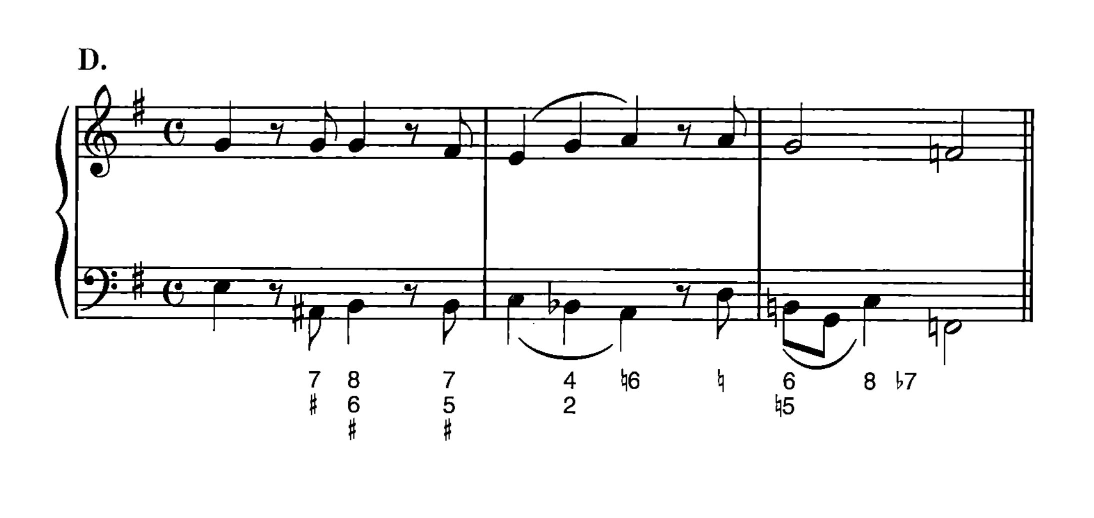

NAME ______________________

2 Compose three original passages, each of which employs the indicated foreign modulation. Use the indicated modulatory technique.

A. B♭ major to VI (G major), by common tone.

B. E major to ♭III (G major), by change of mode.

C. A♭ major to VII (G major), by altered pivot chord.

3 Provide a roman-numeral analysis of the following excerpts, indicating the original key, the new key, and the means of modulation.

A. Schubert's song "Die Sterne" explores a number of foreign keys related by 3rd to the original tonic of E♭ major. What scale degree is the common tone used to link the two keys in each passage?

NAME ____________________

B. The key signature here may be misleading, as it actually refers to the original tonic of F minor (the clarinet part is notated at concert pitch). Determine the key center in measures 1–6 and then the new key in measures 7–8 before determining the means of modulation.

BRAHMS: CLARINET SONATA NO. 1 IN F MINOR, OP. 120, NO. 1

C. Notice that Verdi notates the new key enharmonically. Is there a common tone between the two keys?

VERDI: TRIUMPHAL MARCH FROM *AÏDA*, ACT I

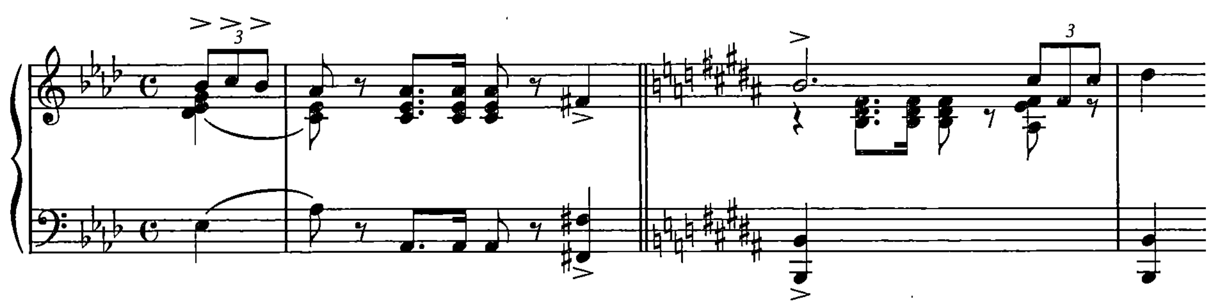

D. The passage cited from this popular song forms a modulatory link between its chorus (mm. 1–2) and bridge (mm. 3–7) sections. Explain the difference between the original first three measures (No. 1) and a recomposed version of these measures (No. 2).

JEROME KERN: "THE WAY YOU LOOK TONIGHT"

E. This striking passage includes two transient tonicizations. Is either temporary tonic related to the original key of E major? How is the tonal transition back to the original key effected?

BERLIOZ: *HAROLD IN ITALY*, II

NAME ______________________________

CHAPTER 35

Harmonic Sequences III

CHROMATIC ELABORATIONS OF DIATONIC SEQUENCES

1 Each of the following two passages consists of a descending chromatic tetrachord. Make a different harmonic setting of each in four-voice texture, and supply a roman-numeral analysis.

A.

B.

2 Realize the following figured-bass exercises, and give a roman-numeral analysis of each. Remember to reserve roman-numeral functions for the opening and conclusion of the sequential passages; use figured-bass symbols for the sequences themselves. Identify the type of each chromatic sequence.

A.

B.

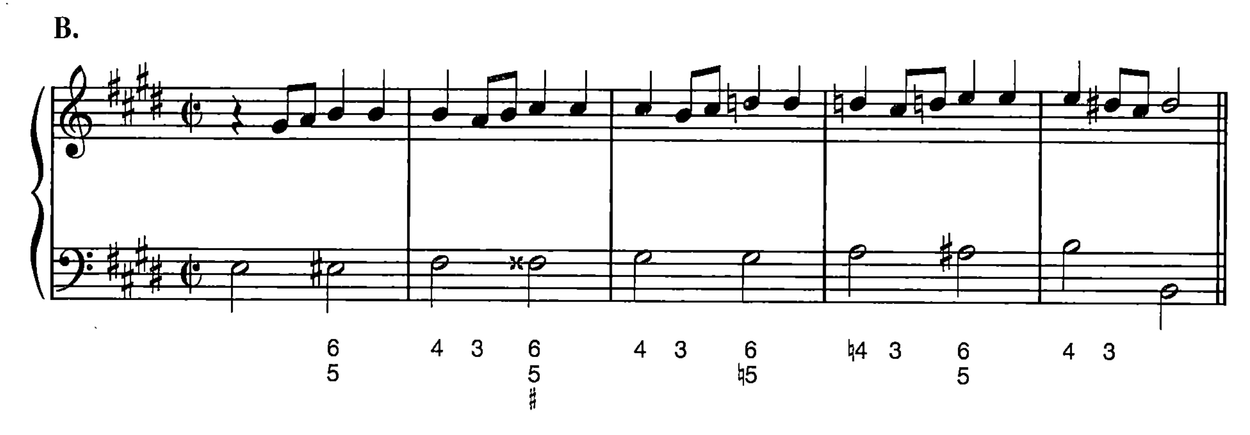

C.

NAME ______________________________

3 For the following two passages, realize the figured-bass symbols in the first one and a half measures and continue the sequence, marked by the brackets, providing a conclusion or cadence for each passage. Analyze the framing chord functions in each passage.

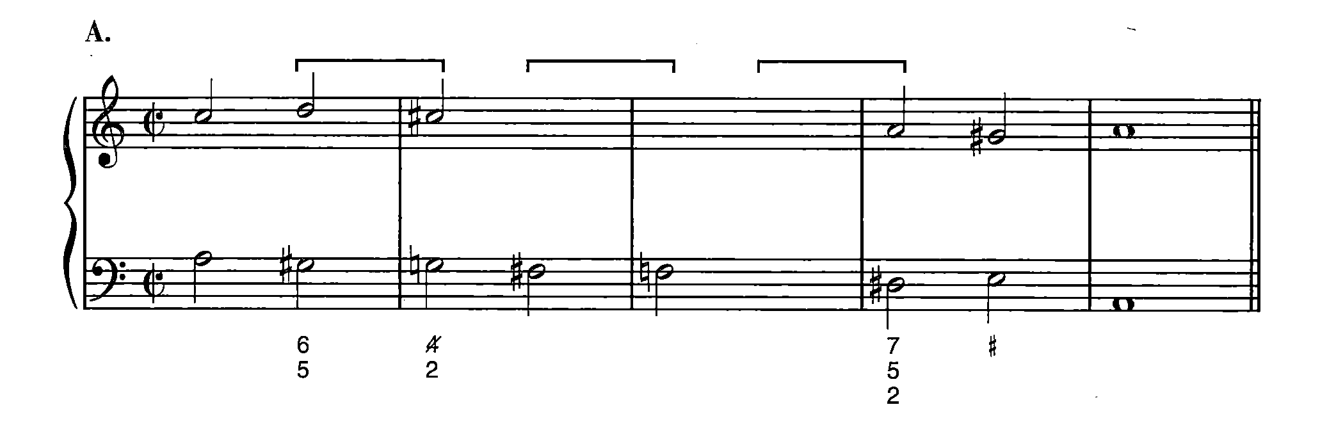

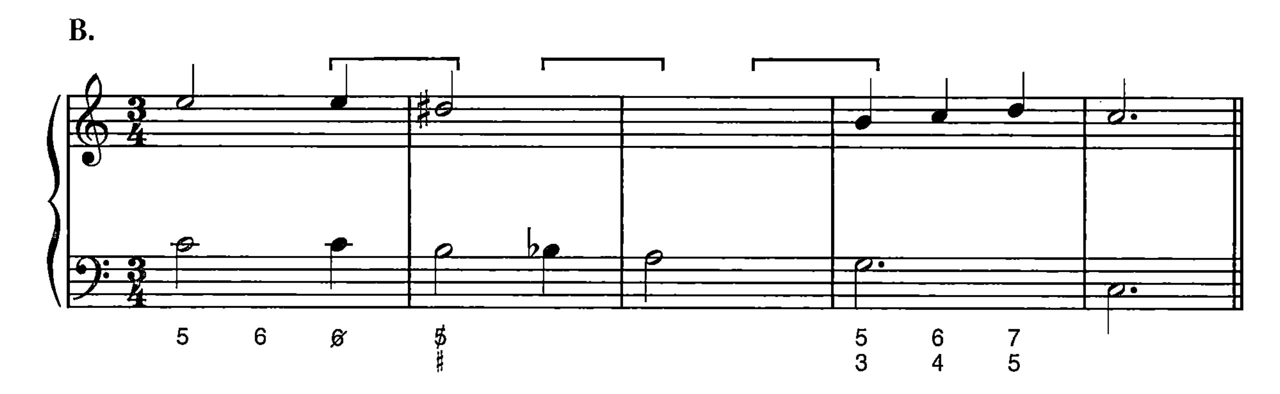

4 Supply a roman-numeral analysis of the following excerpts, identifying each type of chromatic sequence. Provide a voice-leading reduction for those excerpts that have empty staves.

A. This passage is best understood if you read or play it *backwards!*

HANDEL: "O FATAL CONSEQUENCE OF RAGE" FROM *SAUL*, ACT II

B. Why do you think a $\frac{6}{4}$ chord was substituted for the expected dominant harmony in measure 6?

SCHUBERT: STRING QUARTET NO. 15 IN G MAJOR, I

C. Assuming the key of F major, is it feasible to continue this sequence any further? Why or why not?

CORELLI: CONCERTO GROSSO IN F MAJOR, OP. 6, NO. 6, II

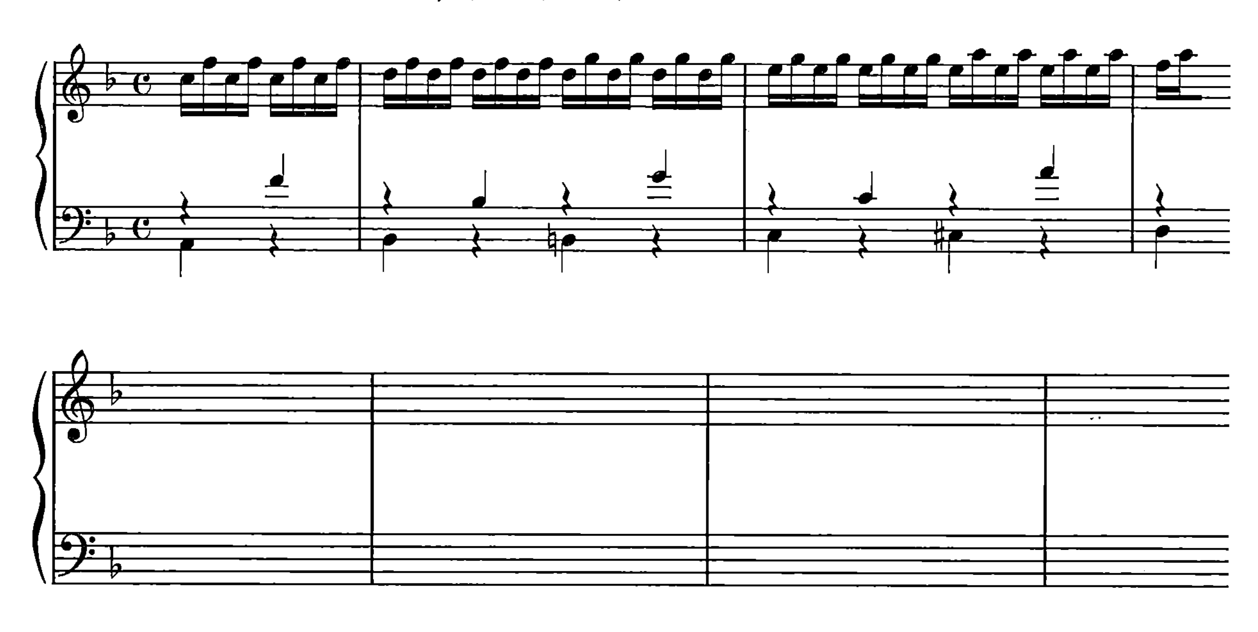

NAME ______________________________

D. What type of triad plays a significant role in this chromatic sequence? Why is the final chord unusual compared to that in measure 3?

CHOPIN: ETUDE IN A♭ MAJOR FROM *TROIS NOUVELLE ÉTUDES*

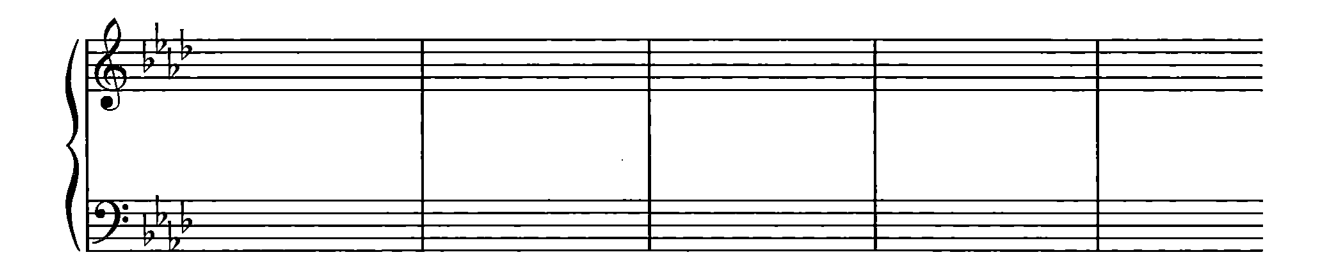

E. In terms of C minor, what is the harmonic function of the last chord? How does it result from the sequential motion?

ALESSANDRO SCARLATTI: FUGUE IN F MINOR

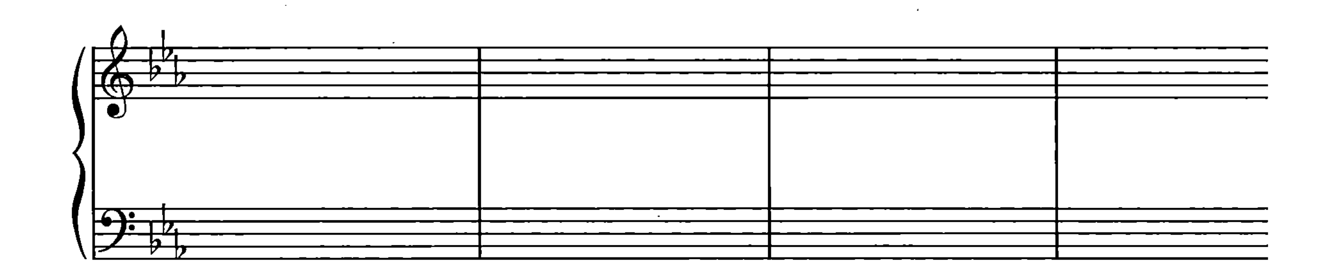

F. What is the root movement between tonicized harmonies? (The passage begins in F♯ minor.) This example is similar to the *Tristan* Love Duet quoted as Example 1 in Chapter 38 of the text.

GOUNOD: LOVE DUET FROM *ROMÉO ET JULIETTE*, ACT IV

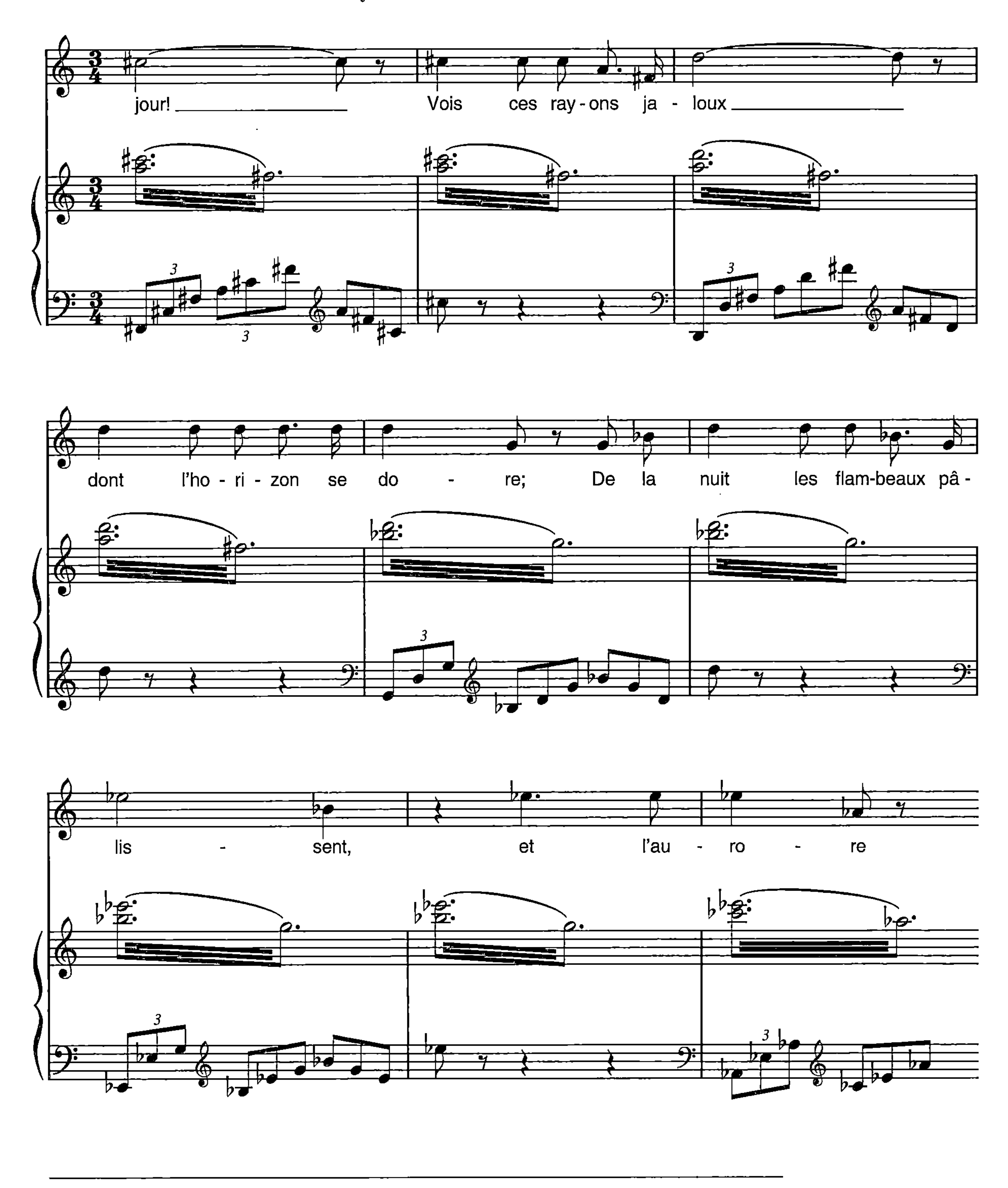

NAME ______________________________

CHAPTER 36

An Analysis Project

(textbook chapter title: Analytical Comments on Wagner's *Tristan* Prelude)

1 The opening section of Isolde's final aria from *Tristan* is quoted below. Although it is more commonly known as the "Liebestod" (or "Love-death"), a title that originated with Franz Liszt, it was entitled "Verklärung" (or "Transfiguration") by Wagner, since this music represents a "spiritual" transformation of the same material found earlier in the ardent Love Duet of Act II.

Before beginning your analysis of this passage, play it over or listen to it on CD3, noting the sense of tonal closure between the chord on the downbeats of measures 1 and 11. What is the basic compositional device that prolongs the A♭ chords of measures 1 and 7, and how is the latter tonally linked to the final A♭ $^6_4$ in measure 11? Trace the stepwise motion in the soprano part that rises from A♭^{4} to A♭^{5} in measures 1–8. What enharmonic progression prepares the key shift to B major?

This music is filled with analytical problems. For instance, is the opening A♭ $^6_4$ an essential tonic triad in second inversion, or is it subordinate to the succeeding E♭ V^7, producing an internal $^6_4$? Some of the harmonies may be difficult to classify by roman-numeral symbols.

There are several aspects of the section that relate to certain features of the *Tristan* Prelude, discussed in Chapter 36 of the text. Can you find a passing allusion to the "Tristan" chord notated enharmonically at the same pitch level? How is the basic root movement in measures 1 and 3 reminiscent of the opening of the Prelude? Observe the pair of enharmonic 4ths (E♭^{4}–A♭ = D♯–G♯ and G♭^{4}–C♭ = F♯–B), which are embedded in measures 2 and 6 of the Prelude. On a larger structural level, what is the logic of the motion to B major in measure 12?

Sehr mässig beginnend.
pp
Mild und lei - se wie er lä-chelt, wie das Au - ge
pp
hold er öff - net, seht ihr, Freun-de, säh't ihr's nicht?
p cresc.
7
Im - mer lich - ter wie er leuch - tet,
3

stern - um - strah - let hoch sich hebt?
p molto cresc.
f
dim.

Etwas bewegter.
Seht ihr's nicht?
p dolce

NOTES

NOTES

PART FOUR

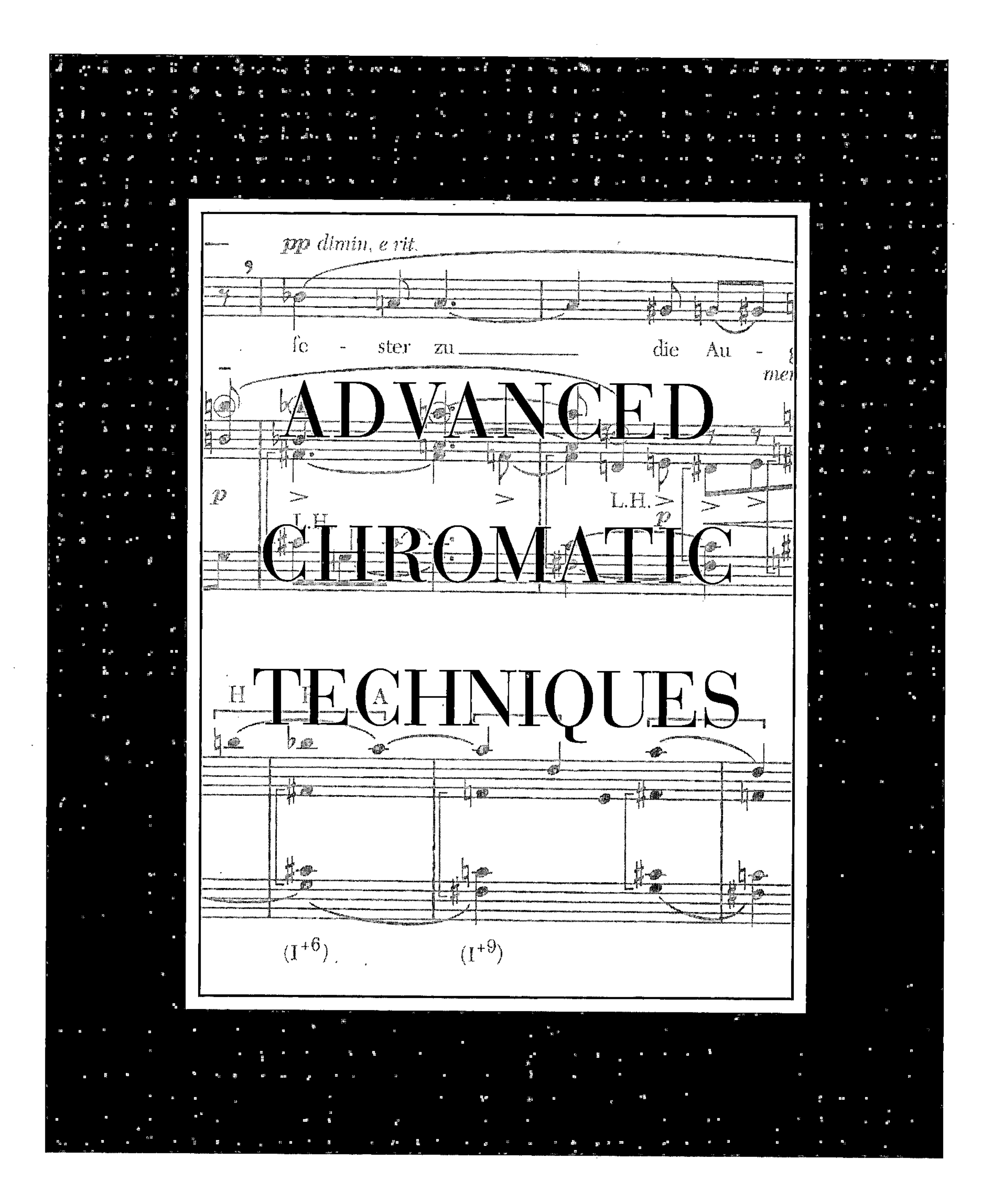

ADVANCED CHROMATIC TECHNIQUES

NAME ______________________

CHAPTER 37

Chromatic Voice Leading

1 Realize the pair of figured-bass exercises below. Give a roman-numeral analysis of Exercise 1A, observing any instances of prolongation by contrary chromatic motion. In the symmetrical chord progression in Exercise 1B, circle the axis chord.

B.

2 Continue the following sequences in strict fashion. What chord type is featured in Exercise 2A? What device is exploited in Exercise 2B?

A. (BASED ON CHOPIN'S MAZURKA IN C♯ MINOR, OP. 30, NO. 4)

B.

Name ______________________________

3 Supply a roman-numeral analysis for each of the following excerpts. You do not need to provide voice-leading reductions.

A. How is the dominant prolonged in this excerpt?

Verdi: Quartet from *Rigoletto*, Act III

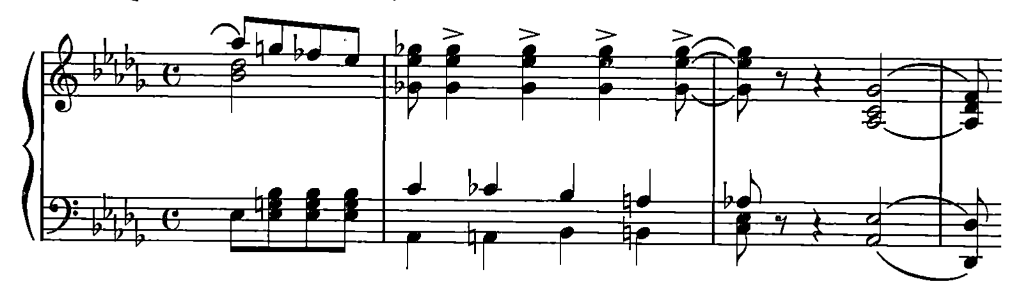

B. This ominous music opens the famous Wolf's Glen scene in Weber's *Der Freischütz.* While the passage exhibits a prominent chromatic descent in the bass, it is largely nonsequential. Mark the small-scale pattern repetition in the middle of the excerpt. Measures 8–10 set up the expectation of tonicizing in what key? Why isn't it possible to assign a roman-numeral function to every chord in this progression?

Weber: Wolf's Glen Scene from *Der Freischütz*, Act II

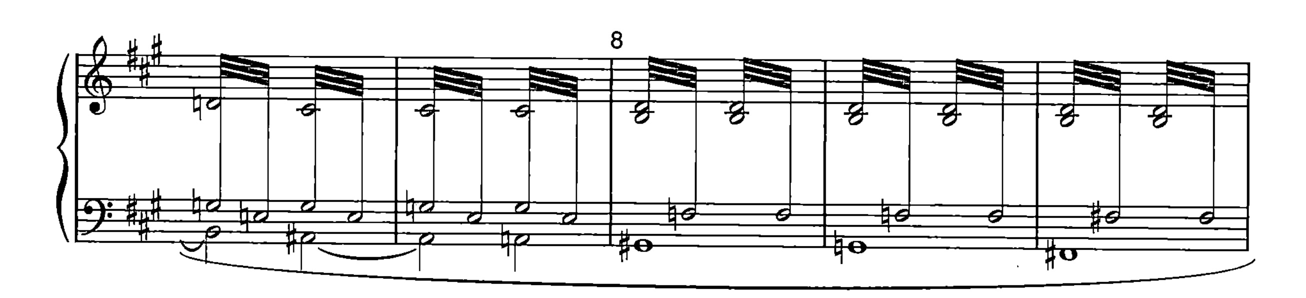

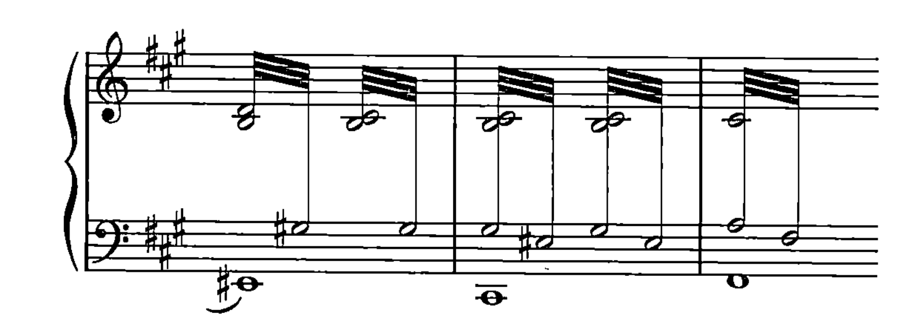

4 Make a voice-leading reduction of this interesting passage. Although contrary chromaticism forms its basis, this motion is not strict throughout. Bracket an instance of a voice exchange and state whether it is diatonic or chromatic.

TCHAIKOVSKY: WALTZ OF THE FLOWERS FROM *THE NUTCRACKER*, ACT II

Name ______________________________

CHAPTER 38

Modulation to Foreign Keys II

1 Realize the following figured-bass exercises. Each example modulates to a foreign key, using either an enharmonic pivot chord or an exact phrase sequence. Your accompanying roman-numeral analysis should indicate how the modulation occurs. In measures 1 and 5 of Exercise 1D, do not begin your four-part texture until beat 2.

A.

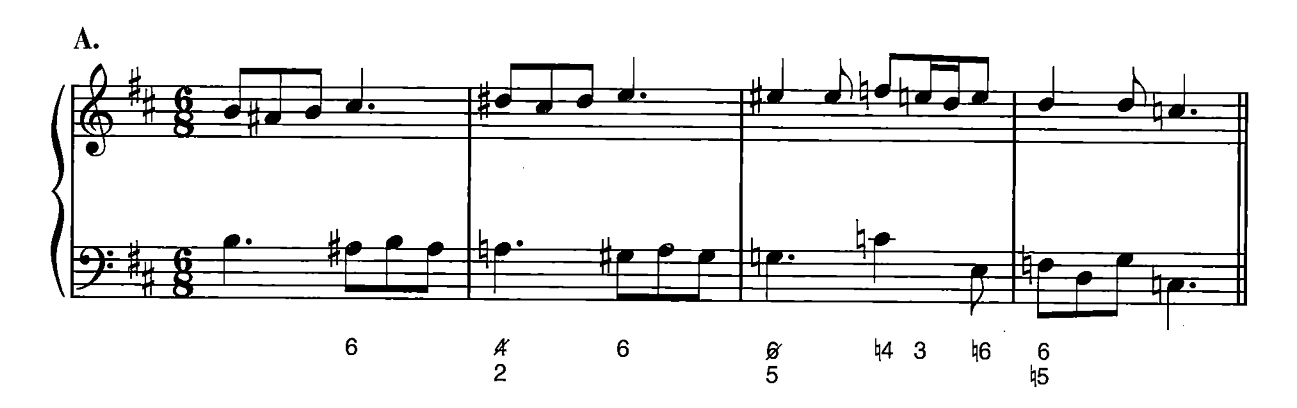

B.

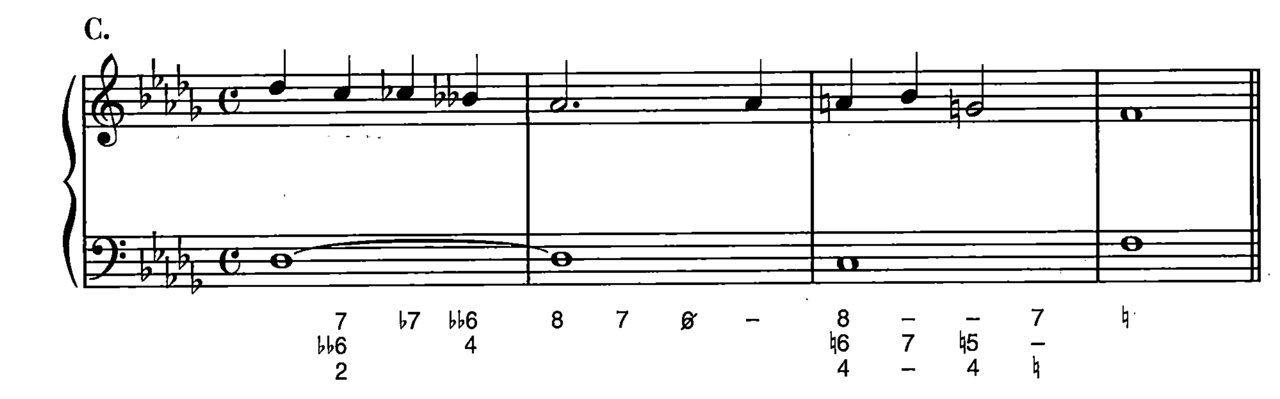

2 Compose three short original passages that utilize the following methods of modulation from and to the given keys.

A. D major to F♯ major, using an enharmonic G♯ B D F diminished-seventh chord as the pivot.

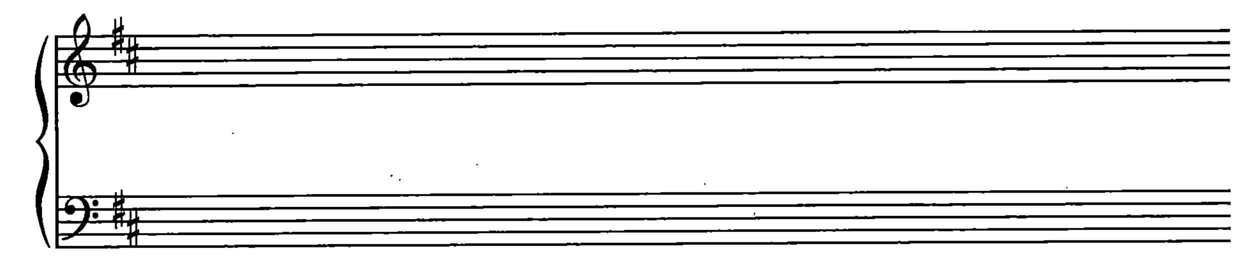

B. F major to E minor, using an enharmonic V^7 = Ger$^6_5$ as the pivot.

NAME ____________________

C. G major to E♭ major, using chromatic voice leading.

3 Supply roman-numeral analyses for the following excerpts. Provide voice-leading reductions for Exercise 3C and 3F. Be sure to indicate in your analysis how the foreign modulation takes place. Although the first three quotations exhibit tonal shifts by ascending or descending half step, the modulatory technique for each is different.

A. What is the crucial chord that introduces the new key? Where do you expect it to go? What actually happens?

MOZART: SYMPHONY NO. 40 IN G MINOR, K.550, I

B. What are the two modulatory procedures that link the key of C major at the opening of this passage to the new key of D♭ major at its conclusion?

BEETHOVEN: SYMPHONY NO. 1 IN C MAJOR, II

C. In this fantasia-like movement, Haydn has dispensed with the usual key signature and instead has written in all of the necessary accidentals. How do you explain the curious notation in measures 5–6?

HAYDN: STRING QUARTET IN E♭ MAJOR, OP. 76, NO. 6, II

D. The opening tonal center of this excerpt is D♭ major. With what harmony does it modulate back to the original F-major tonic? Which other excerpt in Exercise 3 shows a similar function?

BRAHMS: *EIN DEUTSCHES REQUIEM*, I

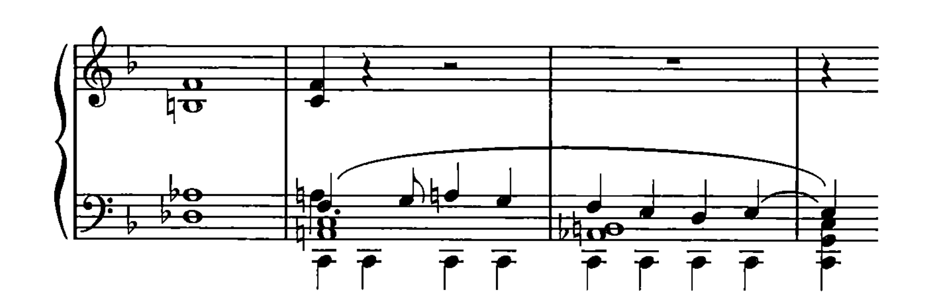

E. The following passage forms a transition to the coda, linking VI–V in F minor. There are several transitory tonicizations that occur in rapid succession, including one to a very remote region. Indicate these in your roman-numeral analysis.

CHOPIN: BALLADE IN F MINOR, OP. 52

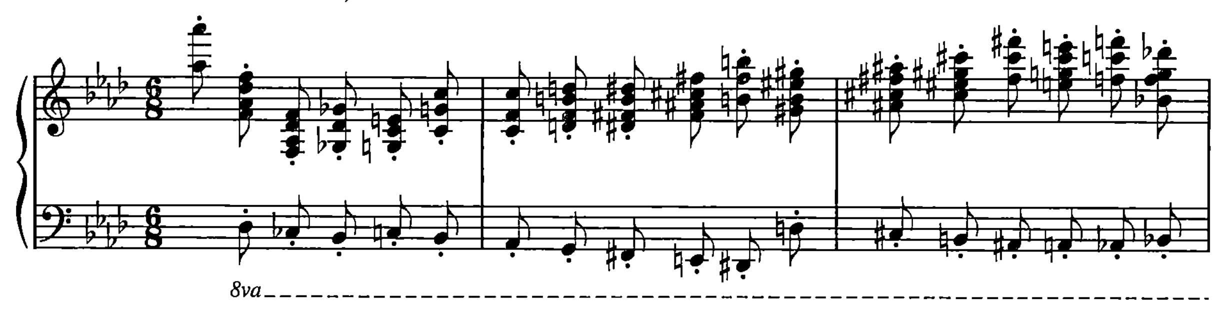

F. This excerpt opens in F minor. What key is suggested in the last measure? What is its relationship to F minor?

BACH(?): "HARMONIC LABYRINTH"

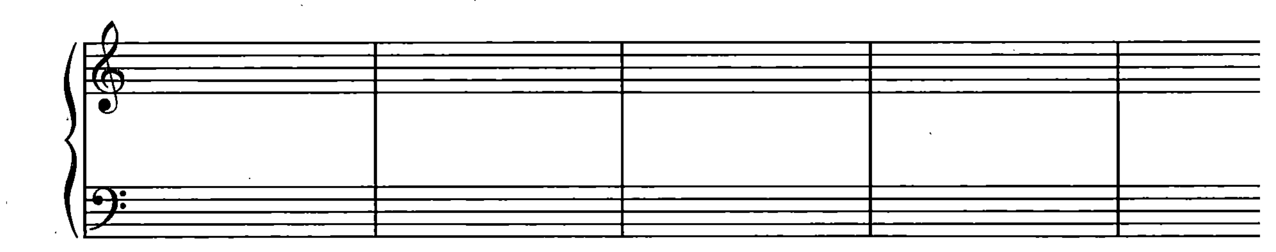

G. Where is this passage structurally divided? What tonal relations exist between the two separate phrases?

FRANCK: VIOLIN SONATA IN A MAJOR, I

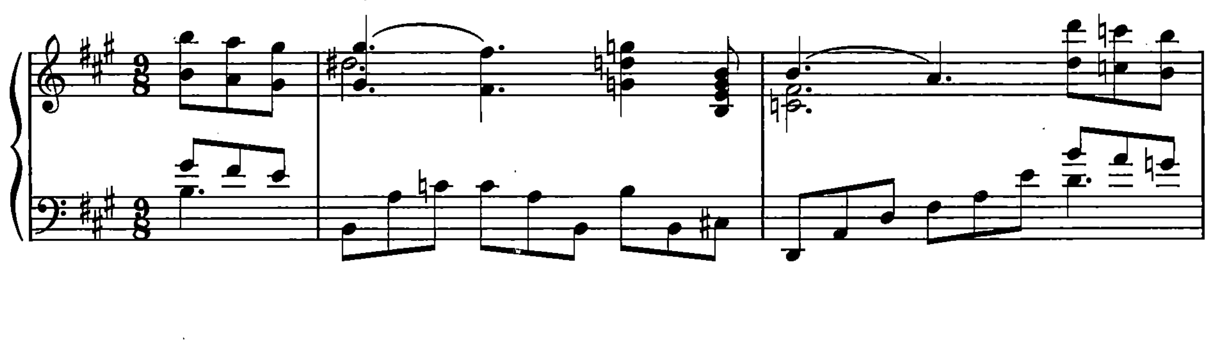

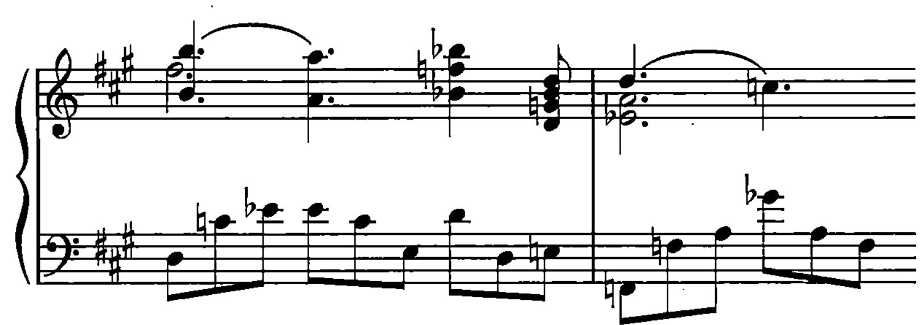

NAME ______________________________

CHAPTER 39

Symmetrical Divisions of the Octave

In this chapter we will concentrate on completing various sequences that employ symmetrical root relations (by major 2nd, minor 3rd, or major 3rd).

1 In Exercises 1A and 1B, continue the bracketed pattern in exact fashion. Indicate what type of root movement is illustrated in each case. In Exercise 1C, continue the second measure to create an omnibus progression by descending minor 3rds. The half-step motion in the upper parts will switch from voice to voice. (Consult Examples 11 and 12 of Chapter 39 in the text.)

A.

B. (BASED ON CHOPIN'S MAZURKA IN A♭ MAJOR, OP. 59, NO. 2)

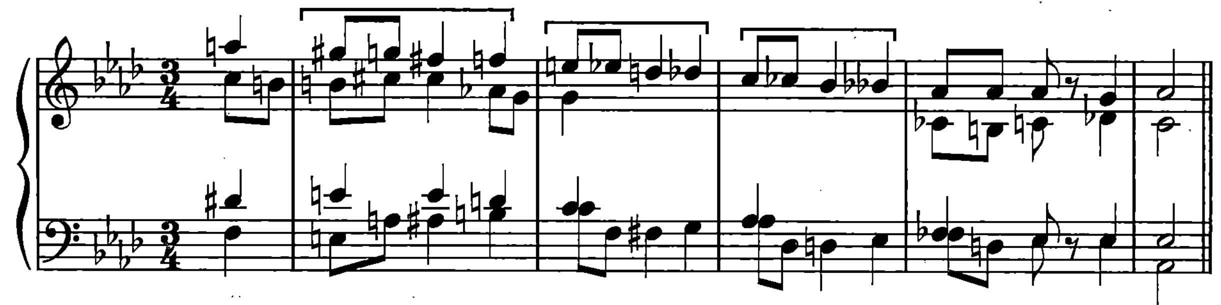

C.

2 The following excerpt is a simplification of a transitional passage from the Scherzo of Beethoven's Symphony No. 9 ("Choral") in D minor. Chart the successive root movements from measure to measure on the blank staff that is provided. What is the intervallic relation between successive pairs of measures? How many of these latter intervals do we traverse?

BEETHOVEN: SYMPHONY NO. 9 IN D MINOR ("CHORAL"), II

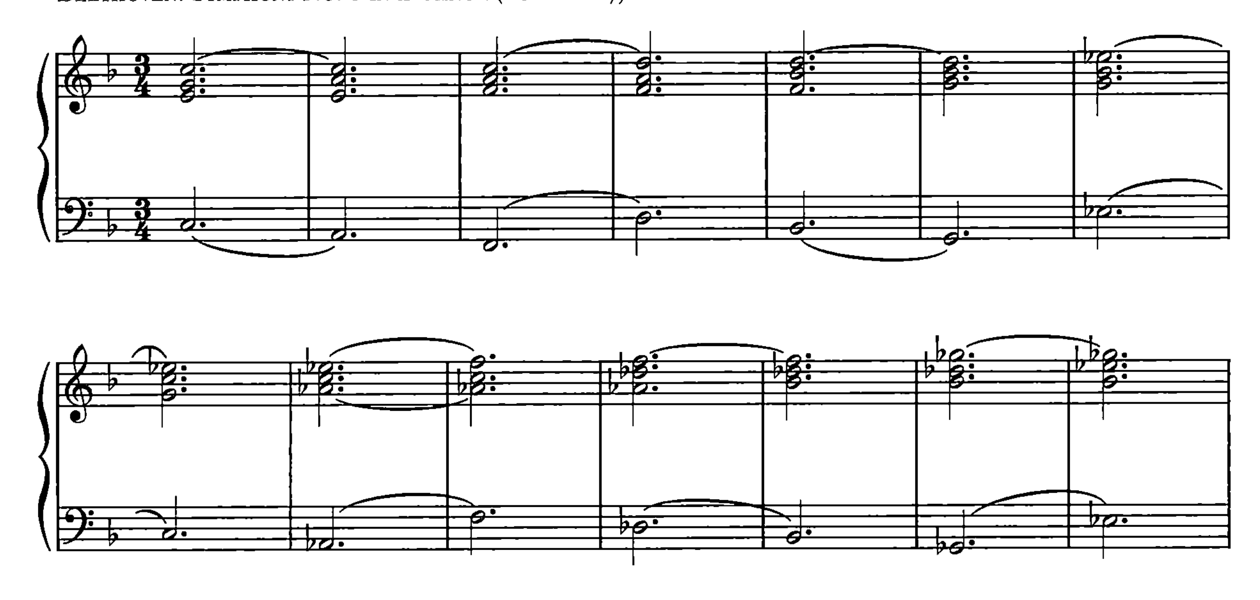

NAME ______________________________

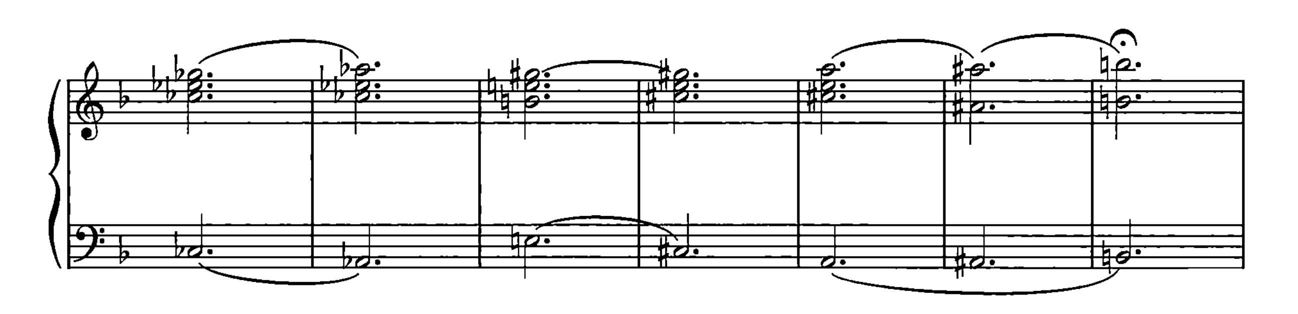

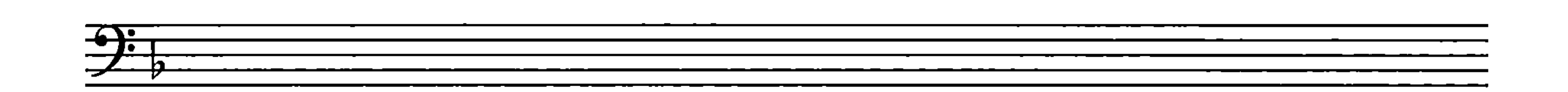

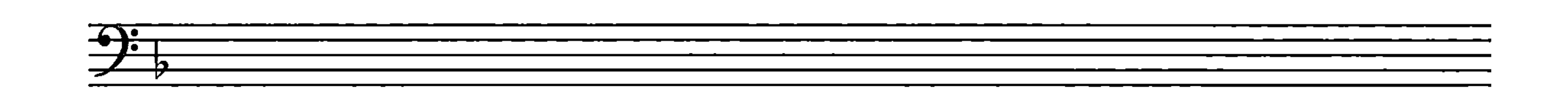

3 Compose an original passage for piano, using a harmonic progression that symmetrically divides the octave. Try to link the basic root movement with additional chords, as in the passages given in Exercise 1.

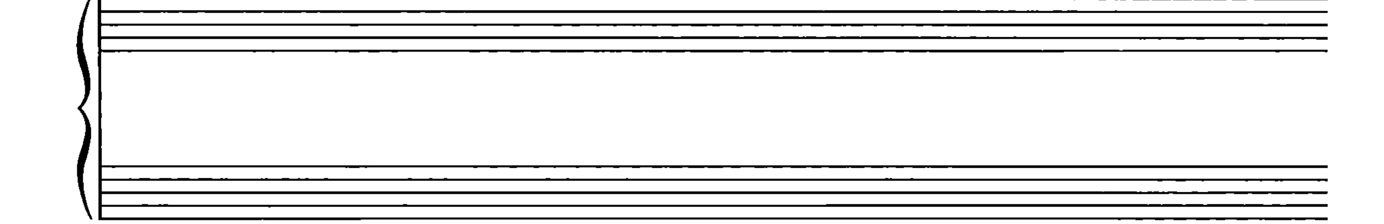

4 Supply roman-numeral analyses for the following excerpts, noting the type of sequence and root movement that are used.

A. What two seventh-chord types are employed in this sequence?

WAGNER: IMMOLATION SCENE FROM GÖTTERDÄMMERUNG (VOCAL PART OMITTED)

B. The commercial chord symbols to a portion of John Coltrane's "Giant Steps" (starting in measure 8) are given below. Indicate the overall root movement of this quotation. What local diatonic progression occurs within each two-measure segment?

JOHN COLTRANE: "GIANT STEPS"

E♭MAJ7 | Am7 D7 | GMAJ7 | C♯m7 F♯7 | BMAJ7 | Fm7 B♭7 | E♭MAJ7 |

NAME ______________________________

C. This excerpt provides a modulatory link to the following waltz in B♭ major. What type of sequence occurs in measures 2–7?

TCHAIKOVSKY: *SLEEPING BEAUTY*, ACT I

D. This strange passage was actually composed in the Baroque period! Chart the sequential root movement in the empty staff provided. How is this excerpt similar to the Beethoven excerpt in Exercise 2? What is remarkable about the partwriting connections between consecutive chords?

DOMENICO SCARLATTI: SONATA IN E MINOR, K. 394

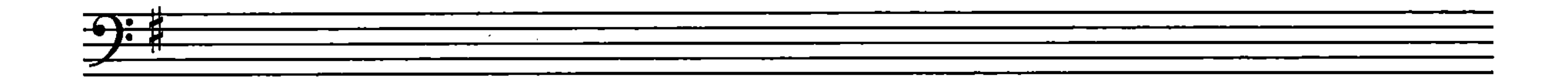

E. How can you justify the statement that this passage from Chopin's D♭ Nocturne represents a "dominant prolongation"? Note the common tone F♯ = G♭ in the upper voices. How does it relate to the underlying chordal sequence?

CHOPIN: NOCTURNE IN D♭ MAJOR, OP. 27, NO. 2 (REDUCTION OF MM. 40–46)

F. Here the rising chromaticism occurs in the bass voice; compare to Exercise 4B. In the upper voices, what is peculiar about the chord connections in this chromatic progression? Can you reposition these parts using octave displacement to produce more standard voice leading? What two keys does this passage link?

BEETHOVEN: STRING QUARTET IN B♭ MAJOR, OP. 18, NO. 6

G. This excerpt contains a number of different exotic chords and compositional devices. Identify them on the score. What is the eventual tonal destination?

FRANCK: SYMPHONY IN D MINOR, III

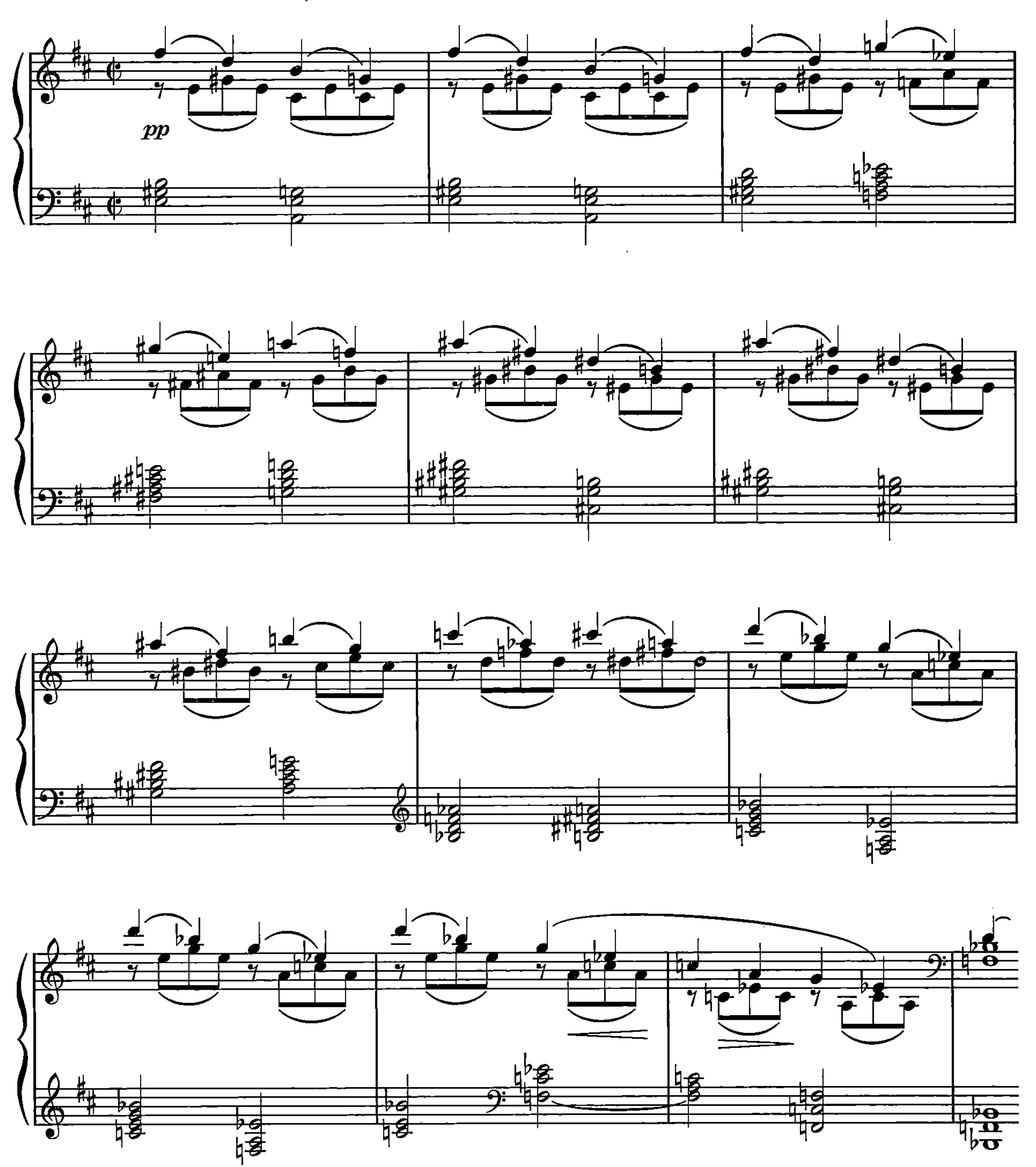

KEYBOARD EXERCISES

CHAPTER 1

Pitch and Intervals

1 Using the given tones in both treble and bass clefs, play the indicated interval above *and* below each note. Recite the pitch you have played with its proper octave designation (F♯[4], C[2], E♭[3], etc.).

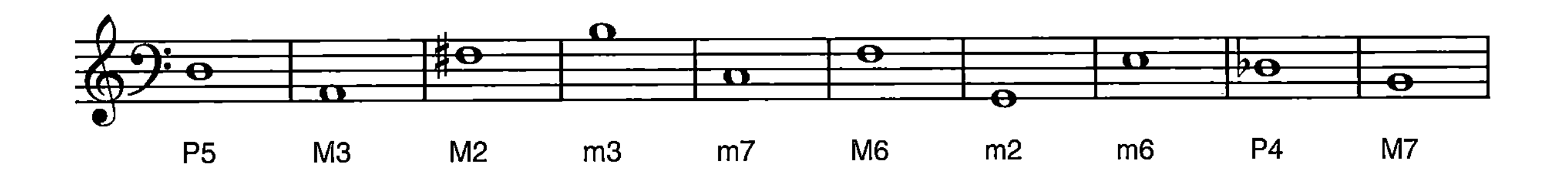

2 Analyze the bracketed interval successions. You may wish to write the names of the intervals into the music. Then continue the intervallic patterns in strict fashion until you reach the final tone that is given. Do *not* write in the notes of the pattern you are continuing.

C.

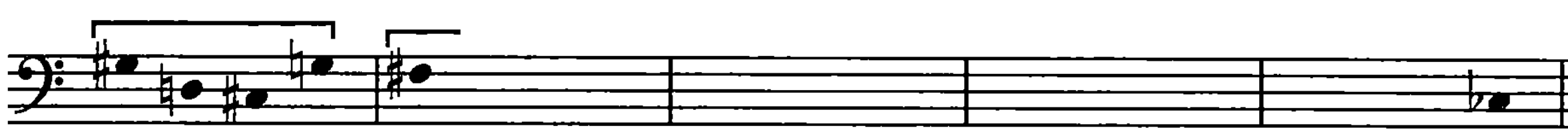

D.

3 Play all the indicated augmented or diminished intervals using each pitch given below in turn. Build the interval upward or downward as indicated by the arrow; note the octave designation.

A. Augmented prime, augmented 5th, augmented 2nd, augmented 4th, augmented 6th on: F⁴ ↑, F♯⁵ ↓, A♭² ↑, D♯³ ↓

B. Diminished octave, diminished 4th, diminished 7th, diminished 5th on: C♯² ↑, B³ ↑, A♯⁴ ↑; D⁵ ↓

CHAPTER 3

Tonic, Scale, and Melody

For the four exercises in this chapter, use the major and minor keys up to four sharps and flats.

1 Play the major scale and the three forms of minor scale (melodic, harmonic, and natural) in all keys up to four sharps and four flats. Recite the interval between successive scale degrees as you play (major 2nd, minor 2nd, augmented 2nd, etc.). Determine the proper key signature of each.

2 Choose a key and play: (1) its relative major or melodic minor scale, and (2) its parallel major or melodic minor scale. Then repeat this procedure for five other keys.

3 Play the indicated scale degrees in several keys.

A. Major: $\hat{1}$ $\hat{6}$ $\hat{3}$ $\hat{5}$ $\hat{7}$ $\hat{4}$ $\hat{2}$ $\hat{6}$ $\hat{4}$ $\hat{7}$ $\hat{2}$ $\hat{5}$ $\hat{1}$

B. Minor: $\hat{1}$ $\hat{4}$ $\flat\hat{6}$ $\hat{2}$ $\sharp\hat{7}$ $\hat{3}$ $\sharp\hat{6}$ $\hat{4}$ $\hat{5}$ $\flat\hat{7}$ $\hat{2}$ $\hat{1}$

4 Using the seven-note rhythmic pattern C ♩ ♩ ♩ ♩ | ♩ ♩ 𝅗𝅥 ‖, improvise melodic phrases in major and minor keys up to three accidentals. Each phrase should begin on $\hat{1}$, $\hat{3}$, or $\hat{5}$ and end with one of the standard melodic cadences (conclusive or inconclusive).

C H A P T E R 4

Triads and Seventh Chords

1 Continue the following pattern of major triads (perfect 5th ↑, perfect 4th ↓) until you reach an F♯ root-position triad. Then reverse the direction, starting from a C[4] triad (perfect 5th ↓, perfect 4th ↑), to cover the flat accidentals. First practice the major-chord succession as shown, and then alter it to a minor-chord succession. You may use one hand only, or you may double in octaves, using both hands.

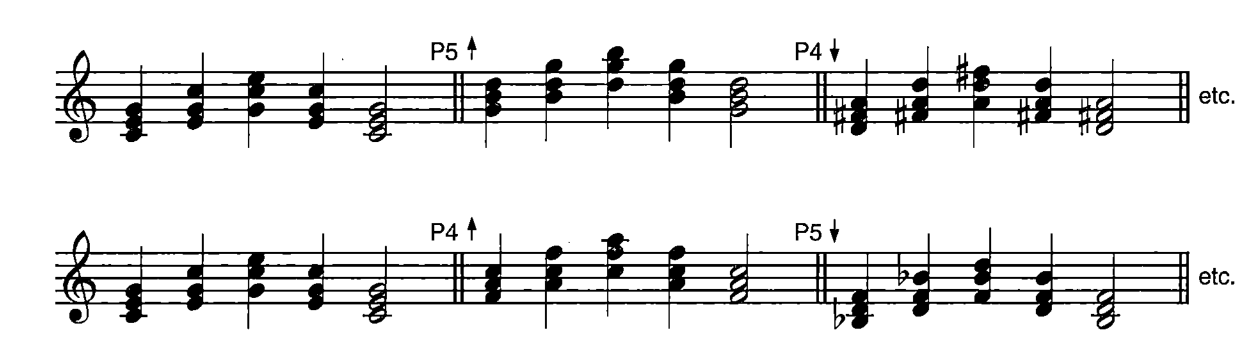

2 Play first-inversion augmented (+) and diminished (°) triads, as indicated, using each of the pitches given below as the chordal *3rd* of the triad. Use the first example as a model.

3 Using your right hand, play the triad indicated by the figured-bass symbols *above* the given tones. Take note of the different key signatures in each example.

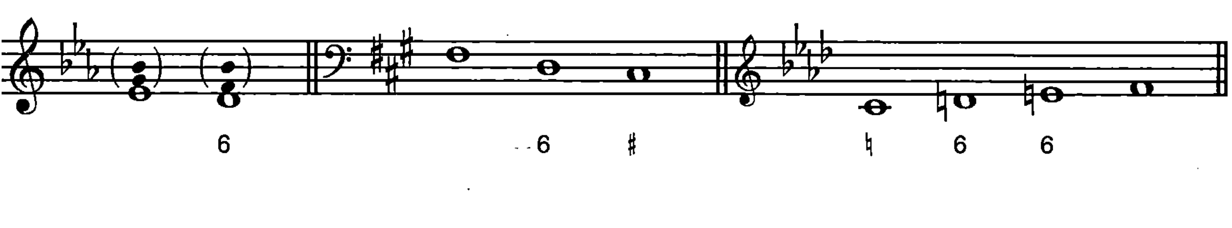

4 Practice constructing all five basic seventh chords (root position only) above the following notes: E, C♯, G, F♯, D, B, A.

5 Continue the chordal patterns given below through at least four more complete measures. Be sure to analyze the sequence of chords before you begin. Identify the root, chord type, and inversion (where applicable) of each chord as you play it.

A.

B.

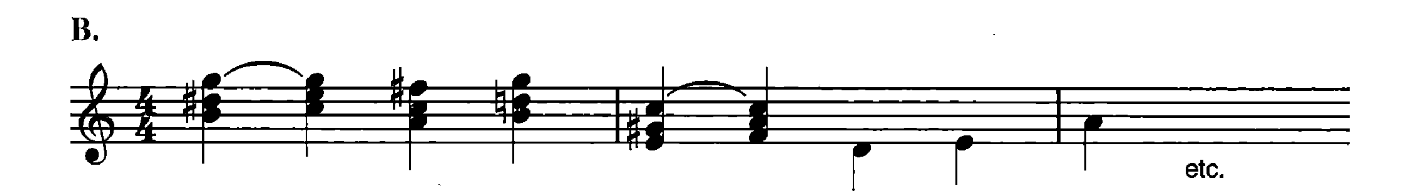

C.

CHAPTER 5

Musical Texture and Chordal Spacing

In the following exercises, realize the figured bass in a four-voice texture, employing *only* close and open/octave structure (in the latter, the soprano and tenor lie exactly an octave apart). Always play the upper three voices with your right hand and the bass with your left hand.

1 Continue Exercises 1A and 1B through the circle of 5ths for at least eight more keys, using the illustrations below as models. Then go back and change the chords to minor triads and repeat the patterns. Repeat these procedures in Exercise 1C, which employs both open/octave and close structures.

A.

B.

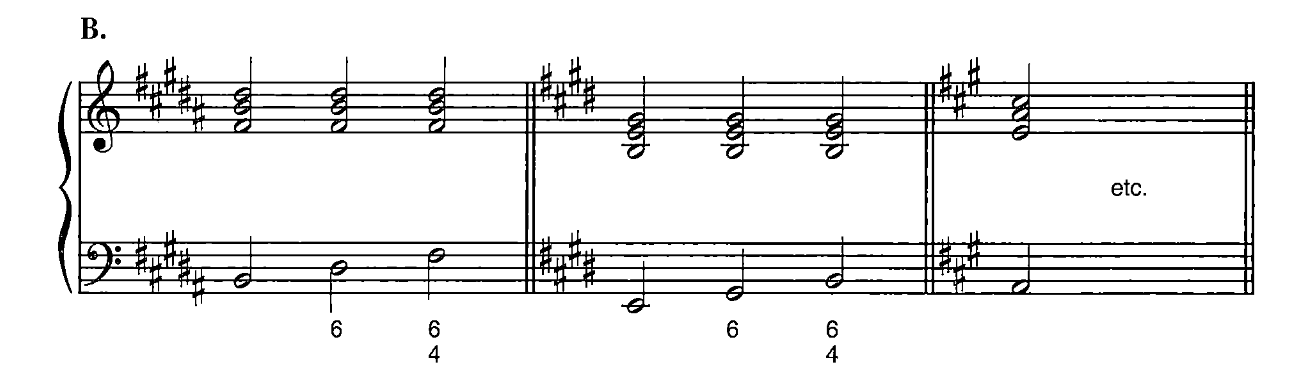

C.

2 Play the triads indicated by the figured-bass symbols in four-voice texture, using either close or open/octave structure. Identify the root, inversion, and chord type of each sonority.

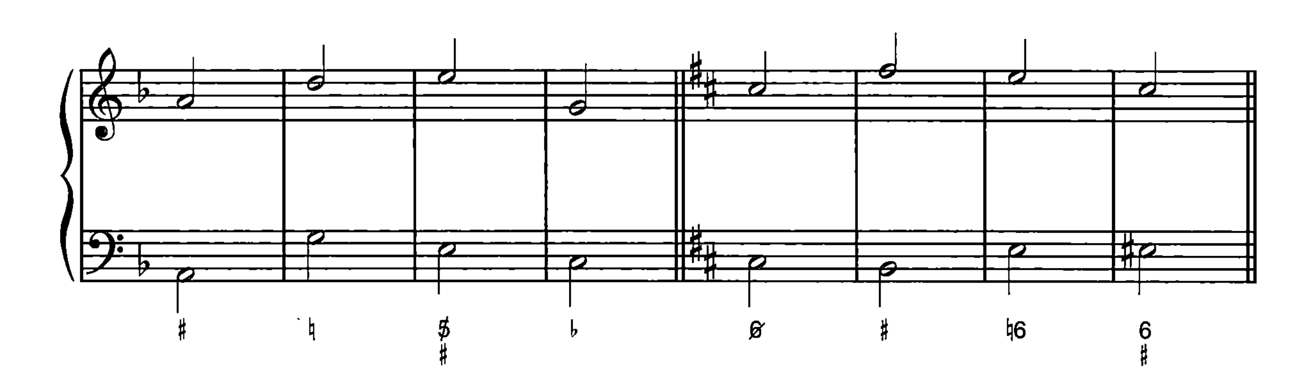

3 Realize the figured-bass exercises given below by filling in the alto and tenor voices. Do not write the notes in the music! Except for the triad marked O/O, play all the chords in close structure. Be sure to observe both the figured-bass symbol and the key signature.

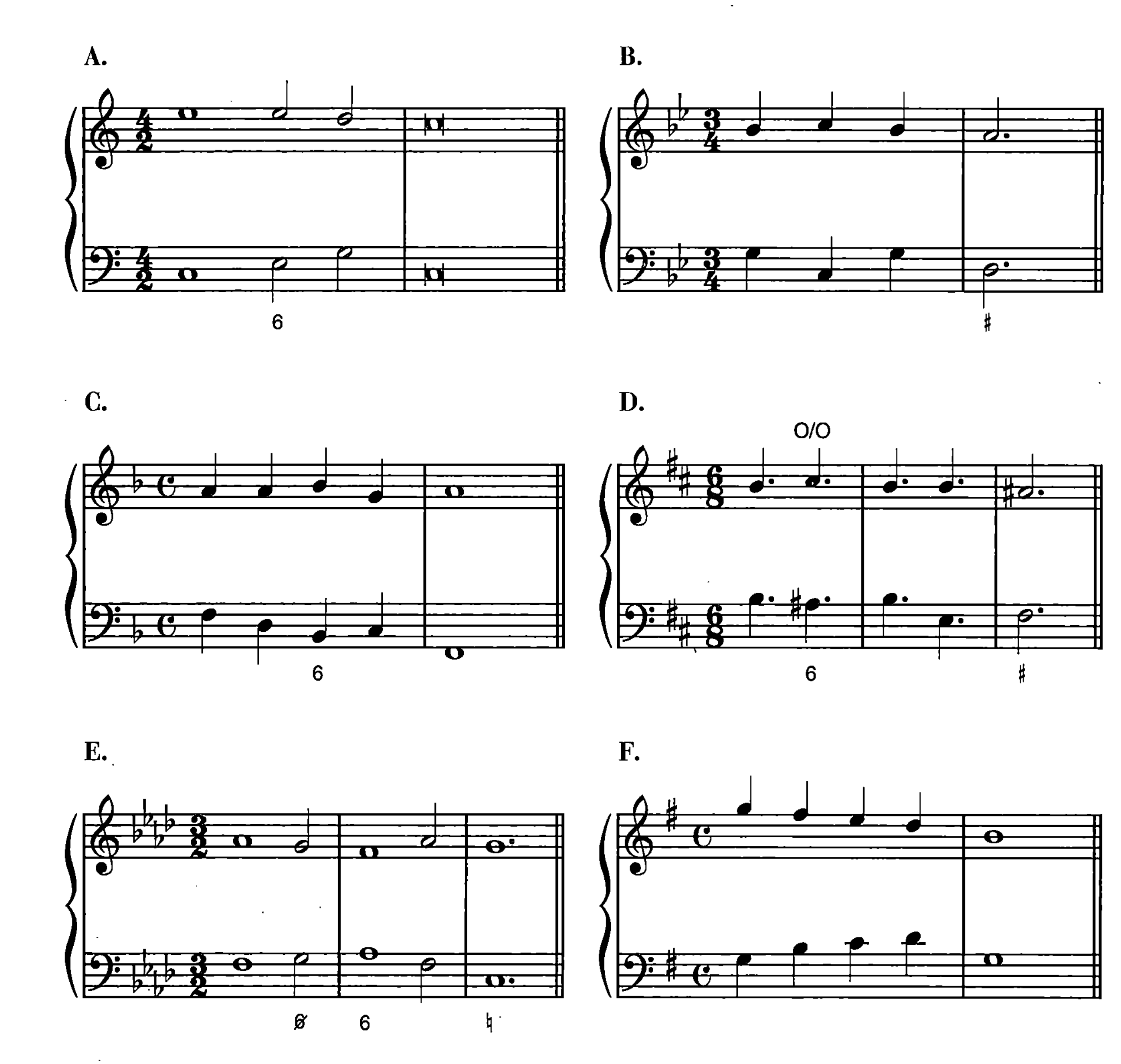

Tonic and Dominant Harmony

1 Play the following two-chord progressions in major and minor keys up to four accidentals; use four-voice texture in close structure. Play all of the triads in root position. The soprano line is given in scale degrees. While the dominant (V) remains a major triad in both modes, the tonic chord will vary from major (I) to minor (i).

A.	$\hat{2}$	$\hat{1}$	**B.**	$\hat{7}$	$\hat{8}$	**C.**	$\hat{2}$	$\hat{3}$	**D.**	$\hat{5}$	$\hat{3}$	**E.**	$\hat{3}$	$\hat{2}$	**F.**	$\hat{8}$	$\hat{7}$
	V	I(i)		V	I(i)		V	I(i)		V	I(i)		I(i)	V		I(i)	V

2 Now improvise six seven-note melodic phrases, each ending with one of the cadential formulas given above. Choose your own major and minor keys. The following two models can serve as examples.

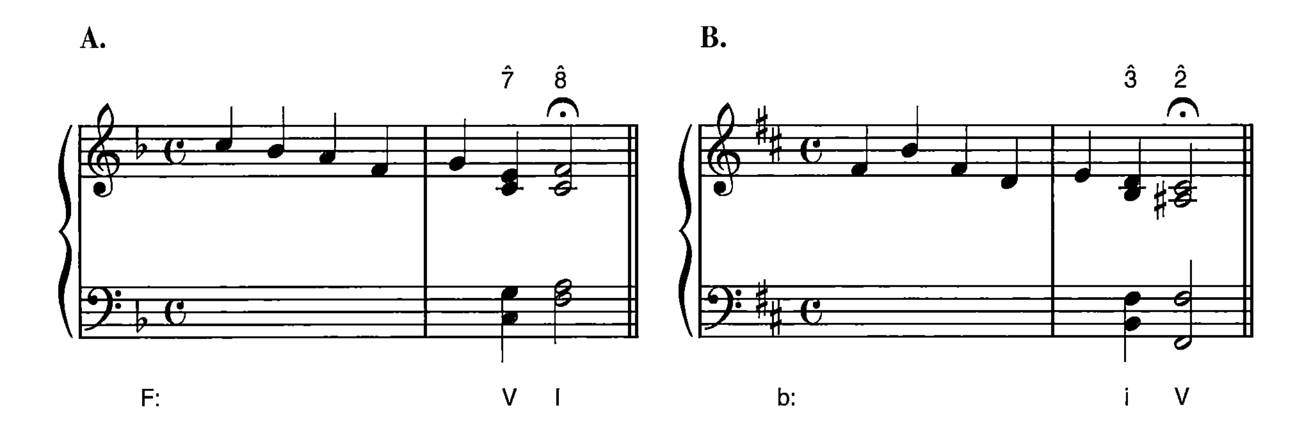

3 Using only close structure, realize the figured bass in the following phrases. Identify the function (cadential or embellishing) of each dominant chord. Watch for the raised 3rds (♮) in Exercise 3B.

A.

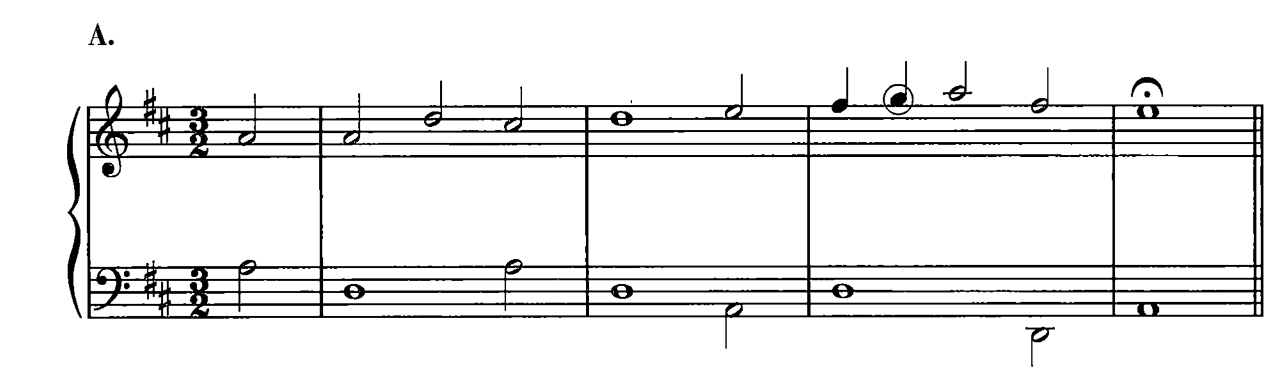

B.

C.

C H A P T E R 1 1

The V^7 and I^6 Chords

1 Practice playing V^7–I(i) in various major and minor keys of your own choice. Use root position for both chords. Be sure that the chordal 7th descends by step to $\hat{3}$.

2 Play each of the following progressions in both C major and C minor, and then transpose them to at least four other major and minor keys of your choice. (The key signatures and accidentals in parentheses, given in this and similar models in later chapters, always represent both C major and C minor.) Observe the approach to and resolution of the chordal 7th in each example. The chordal 5th of the V^7 is omitted in Exercise 2B.

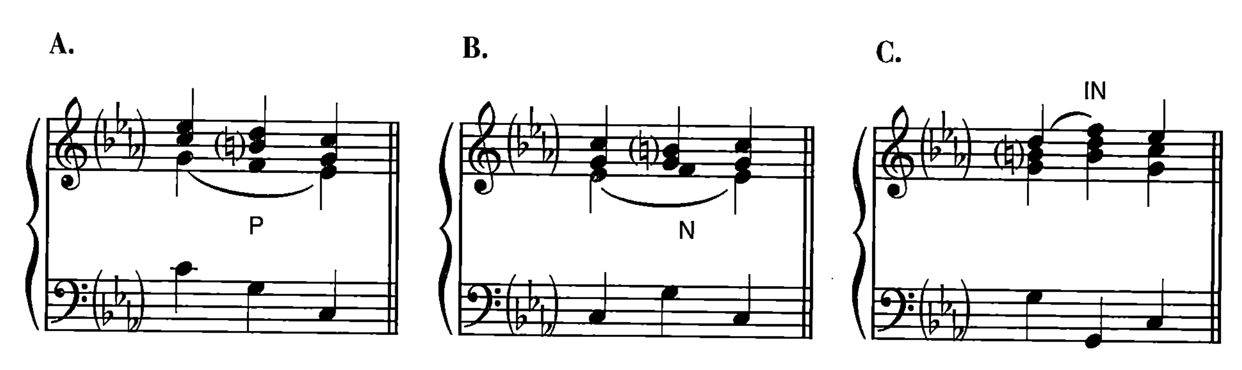

3 Realize the following figured-bass exercises, using close structure for all chords except those indicated by arrows, for which you should use open/octave structure. Make sure that your chord spelling agrees with the figured bass. Explain the treatment of the chordal 7th in the V⁷'s.

4 Harmonize the following two phrases in four-voice texture, utilizing only the I, I⁶, V, and V⁷ chords. Use a V⁷ to support each note marked with an *x*. Make sure that the chordal 7th resolves correctly in the final cadence of Exercise 4B. Remember that using a I⁶ will make your bass line smoother.

CHAPTER 13

Linear Dominant Chords

V^6, $VII^{\circ 6}$, AND INVERSIONS OF V^7

1 Approach (with a I or V chord) and resolve (to I) the various inversions of the dominant seventh: V^7, V^6_5, V^4_3, and V^4_2. Employ a different key for each of these three-chord progressions.

2 The following two progressions contain typical instances of V^6 and $vii^{\circ 6}$, as well as inversions of V^7. Play each progression and then transpose it to keys of at least three accidentals. Be able to give a roman-numeral analysis of each one. For the inverted dominant harmonies, indicate (by P, N, or IN) the melodic derivation of the chordal 7th.

A.

B.

3 Realize the following figured-bass exercises, using four-voice texture and primarily close structure; for the chords denoted by arrows, use open/octave spacing. All nonharmonic tones are circled. Exercise 3C contains a prolongation of the dominant seventh chord.

A.

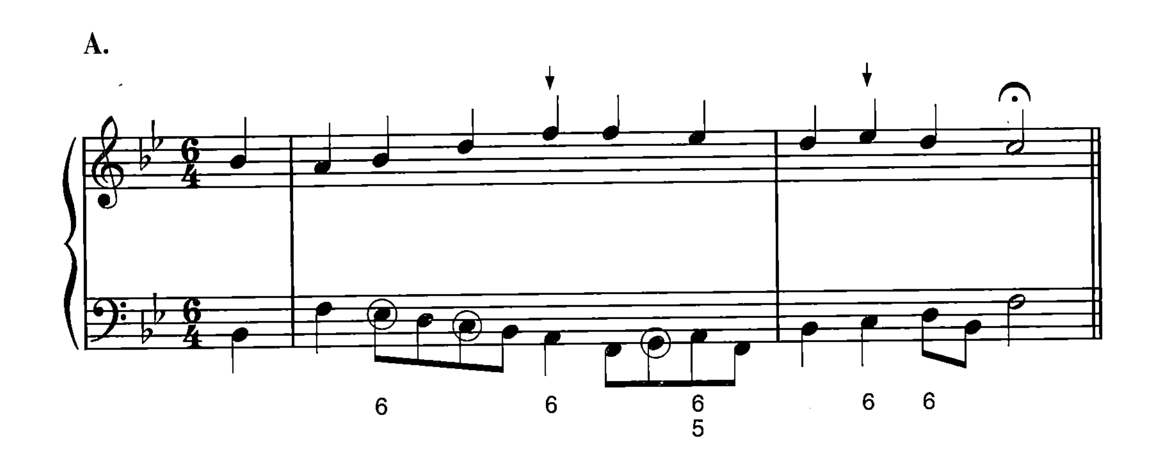

B.

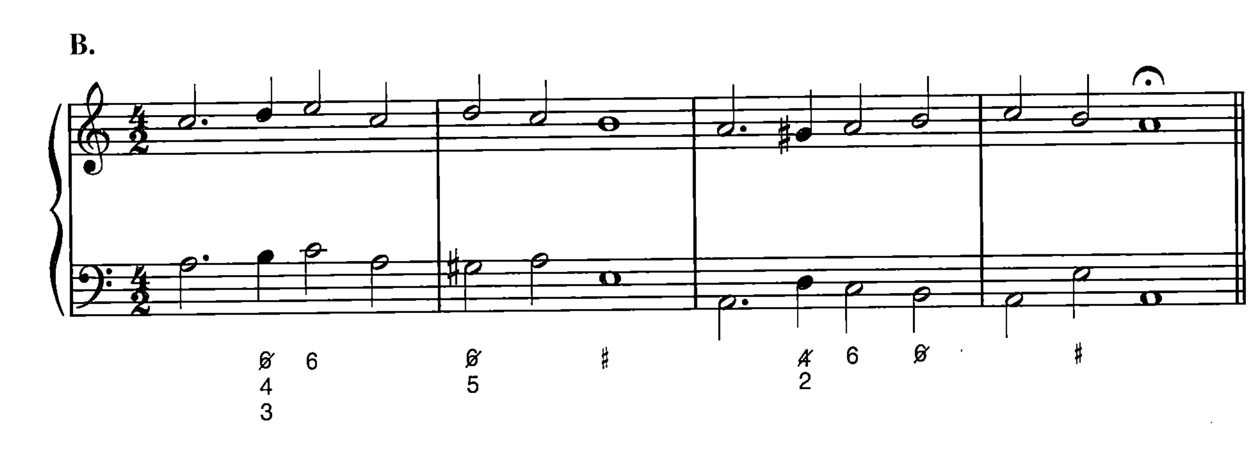

C.

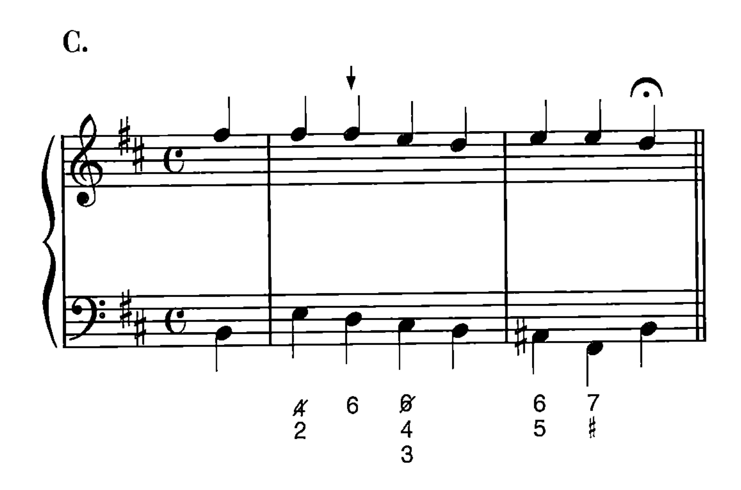

4 Harmonize the following tunes in four-voice texture, using the tonic and dominant-family chords discussed in Chapter 13. Strive for smooth voice leading in your bass lines.

A.

B.

C H A P T E R 1 4

Melodic Figuration and Dissonance II

SUSPENSIONS AND SIMULTANEOUS DISSONANCES

1 Play the following progressions in four-voice texture, using keys of not more than three accidentals. For each progression, the mode (M or m) and scale degrees of the soprano line are given. Place the indicated suspension in the alto or tenor voice; in Exercise 1D, place it in the bass voice. Do not double the suspended note or its resolution in 4–3, 7–6, and 2–3 suspensions; do double the resolution note in the 9–8 suspension.

A. $\hat{5}$ $\hat{3}$ (9–8) $\hat{2}$ (4–3) $\hat{1}$
(M)V^6 I V I

B. $\hat{5}$ $\hat{3}$ (7–6) $\hat{2}$ (4–3)
(m) V i^6 V

C. $\hat{3}$ $\hat{2}$ (7–6) $\hat{1}$
(M) I vii^{o6} I^6

D. $\hat{3}$ $\hat{5}$ $\left(\begin{smallmatrix}5-6\\2-3\end{smallmatrix}\right)$ $\hat{3}$
(m)i V^6 i

2 Play the following chains of suspensions and then transpose them to three other keys (both major and minor) of your choice. Exercise 2B employs only three voices; why are the suspensions necessary? Identify each type of suspension as you play it.

A.

B.

3 Realize the following figured-bass exercises, employing close or open/octave structure. Be sure that you correctly prepare and resolve all suspensions. Locate instances of a change of bass and an ornamented suspension.

A.

B.

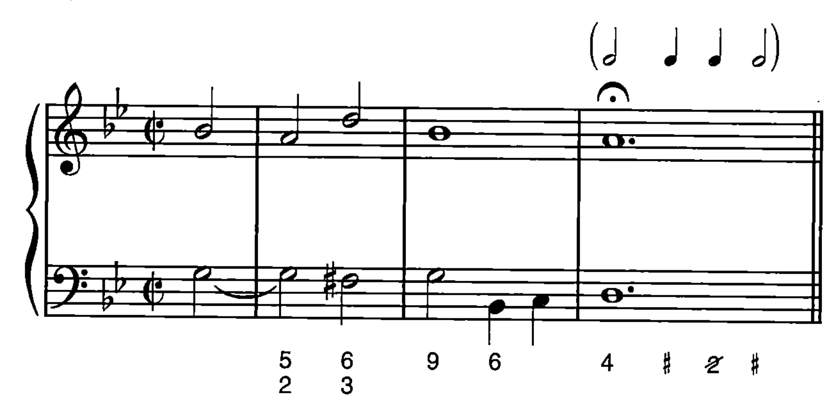

C.

CHAPTER 15

Pre-dominant Chords

IV AND II

1 Play the four cadential formulas given below in keys of not more than three accidentals. Observe the indicated mode and melodic scale degrees. Precede each with three melodic notes to make a seven-note phrase, as in the notated model.

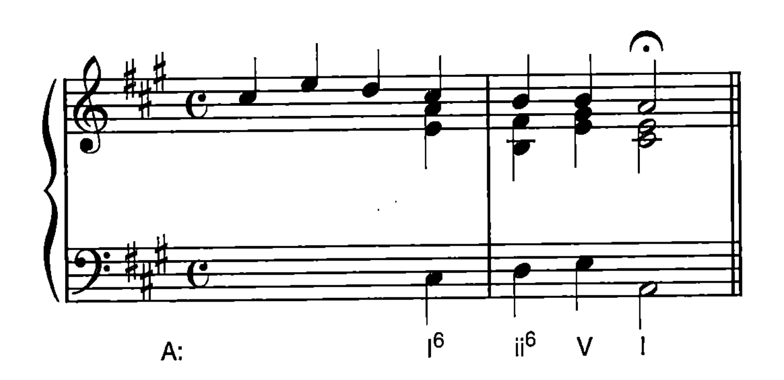

A.	$\hat{8}$	$\hat{8}$	$\hat{7}$	$\hat{8}$	**B.**	$\hat{3}$	$\hat{2}$	$\hat{2}$	$\hat{1}$	**C.**	$\hat{5}$	$\hat{4}$	$\hat{4}$	$\hat{3}$	**D.**	$\hat{3}$	$\hat{2}$	$\hat{7}$	$\hat{8}$
(m)	i^6	iv	V	i	(M)	I	ii^6	V	I	(M)	I^6	ii	V^7	I	(m)	i^6	ii^{o6}	V	i

2 The following progressions represent various ways of embellishing the tonic, using pre-dominant harmonies within the phrase. Transpose each progression to three other keys (major and minor) of your own choice.

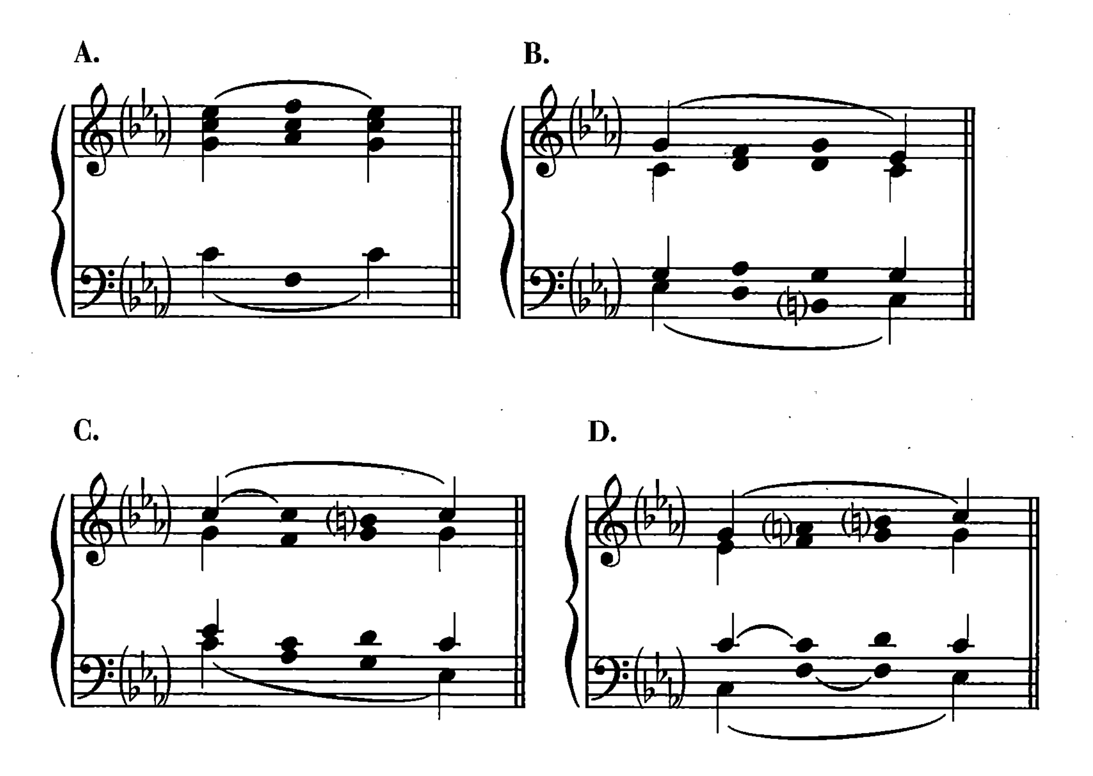

3 Realize the following figured-bass examples, and identify all instances of pre-dominant triads. Remember that in 6 chords, either close or open/octave structure is possible. In close structure, you may double the soprano or an occasional inner voice at the unison.

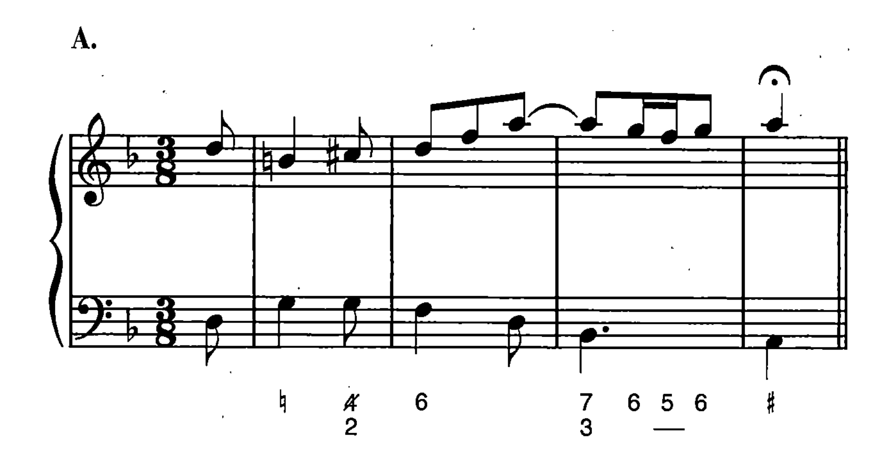

B.

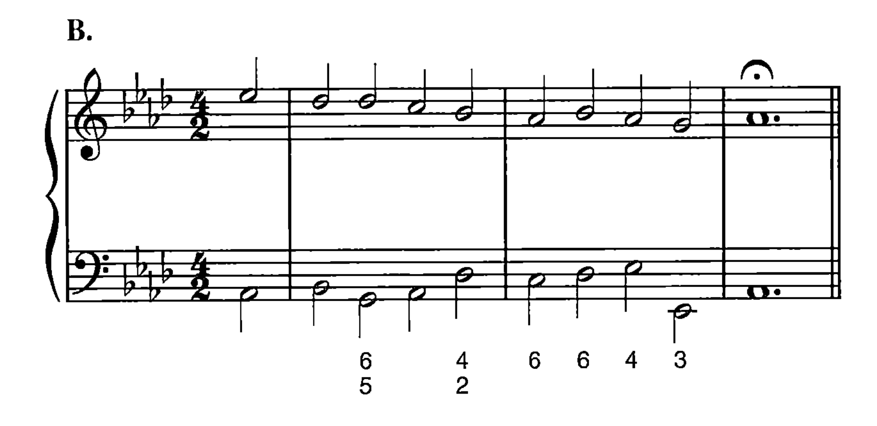

C.

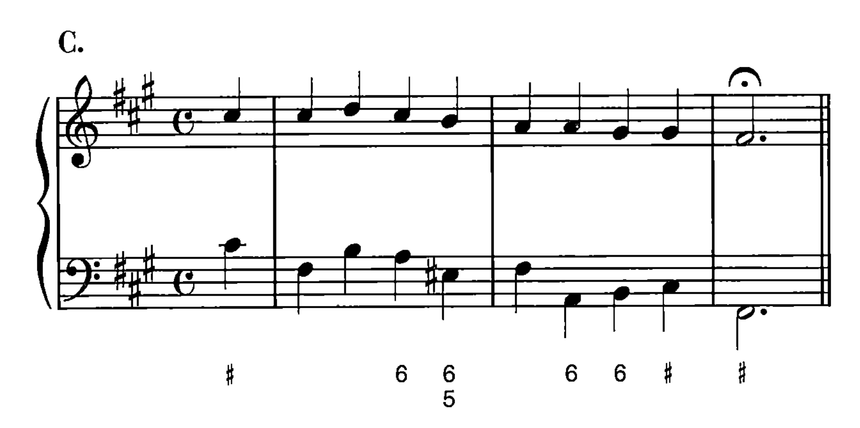

4

Harmonize the following hymn phrases, using appropriate pre-dominant chords in cadential and embellishing roles. What are two ways of approaching the first cadence in "St. Fulbert"?

A. "St. Michael"

B. "St. Fulbert"

CHAPTER 16

The $^{6}_{4}$ and Other Linear Chords

1 The following progressions employ cadential $^{6}_{4}$'s. Play each progression in three different keys of your choice.

A. $\hat{5}$ $\hat{4}$ $\hat{3}$ $\hat{2}$ $\hat{1}$
(M) I^{6} IV V$^{6\ 5}_{4\ 3}$ I

B. $\hat{3}$ $\hat{2}$ $\hat{8}$ $\hat{7}$ $\hat{8}$
(m) i^{6} ii^{o6} V$^{6\ 5}_{4\ 3}$ i

C. $\hat{5}$ $\hat{5}$ $\hat{5}$ $\hat{4}$ $\hat{3}$ $\hat{2}$
(M) I V I^{6} IV V$^{6\ 5}_{4\ 3}$

2 Instances of passing and neighboring $^{6}_{4}$'s occur in the following three models. Play each one and then transpose it to the keys of A, G, and F major and their parallel minors.

A.

B.

3 Realize the following figured-bass examples. Indicate all occurrences of the $^6_4$ and other linear chords that we have studied thus far.

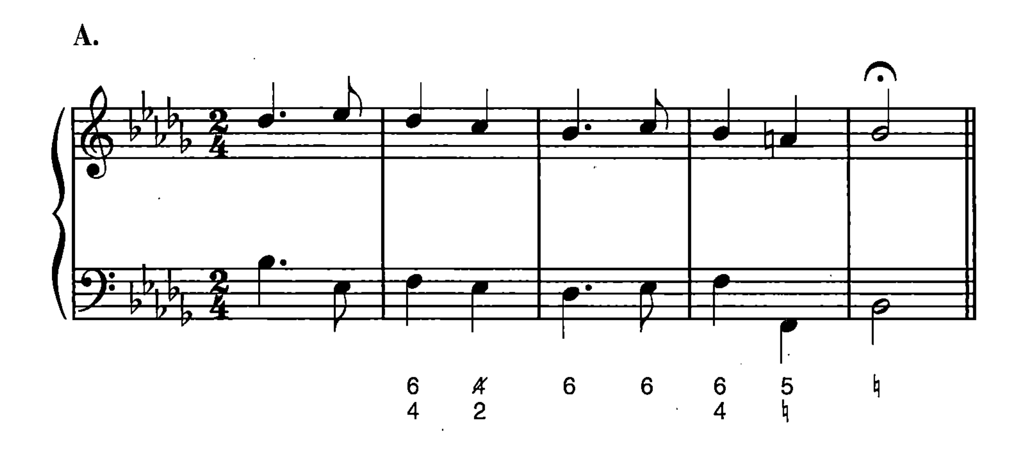

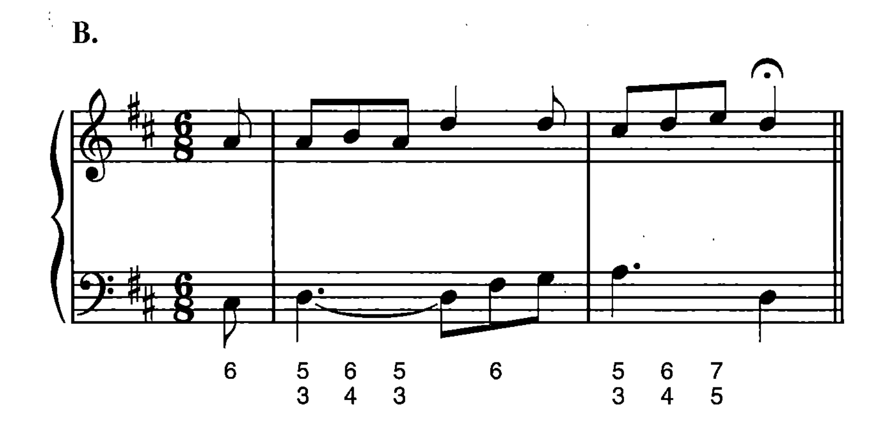

C.

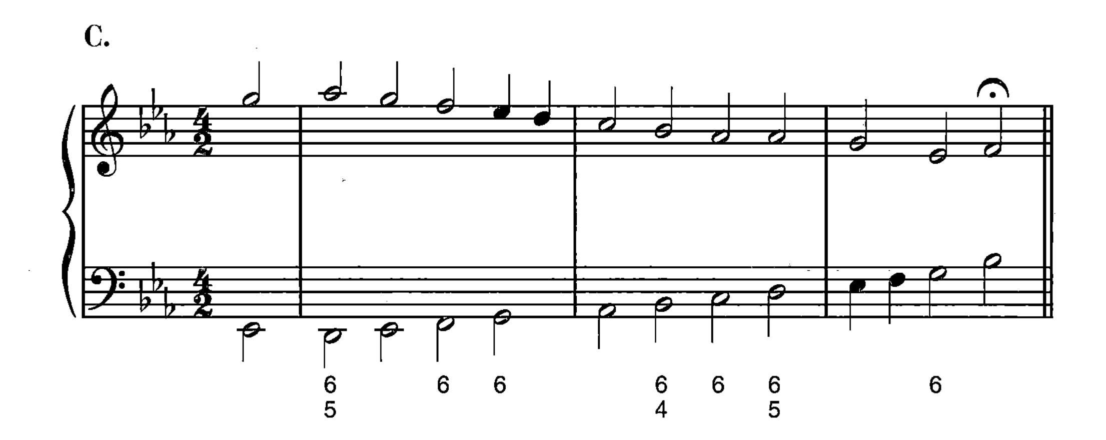

4 Harmonize the following tune in four-voice texture. Place an appropriate $^6_4$ chord at each place marked with an arrow. The labels above the brackets indicate the manner in which the dissonant 4th (which may or may not be the melodic note) is approached and resolved.

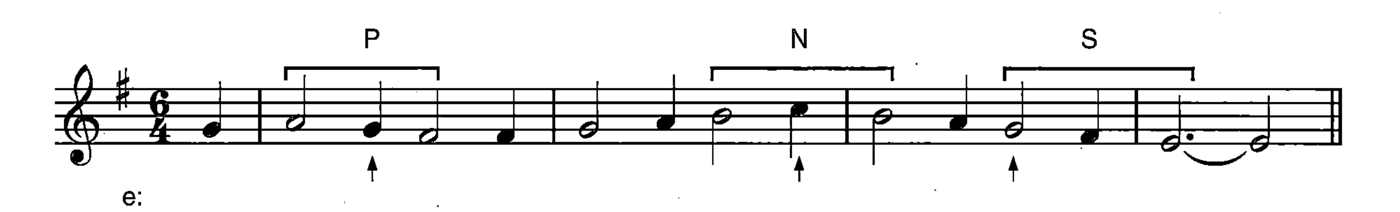

CHAPTER 17

The II^7 and IV^7 Chords

1 Be able to play, in both major and minor keys up to three accidentals, a cadence that contains ii^6_5 ($ii^{ø6}_5$) and/or IV^7 (iv^7). Make sure that the 7th of the pre-dominant chord is prepared (usually with a I^6). The more common soprano lines are $\hat{3}$–$\hat{2}$–$\hat{2}$–$\hat{1}$ (for the ii^6_5) and $\hat{3}$–$\hat{3}$–$\hat{2}$–$\hat{1}$ (for the IV^7). How must you avoid the potential parallel 5ths in the latter?

2 Transpose the following progressions to at least four other major and minor keys of your choice. Locate all instances of pre-dominant seventh chords, and indicate whether they have a cadential or embellishing function.

A.

B.

C.

3 Realize the following figured-bass exercises. Explain the approach to each predominant seventh chord.

A.

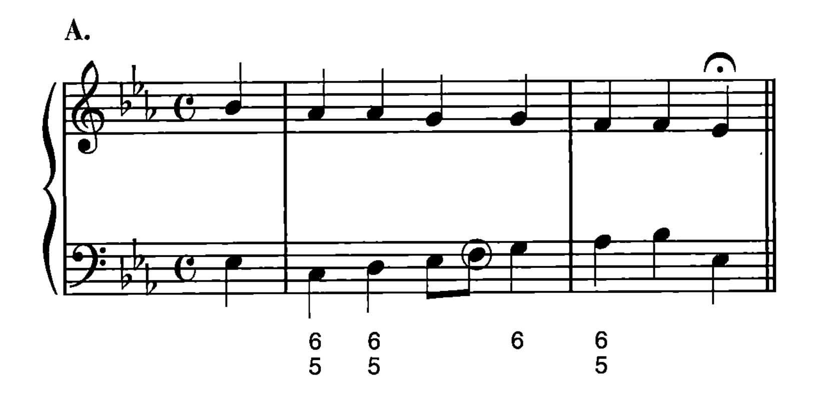

B.

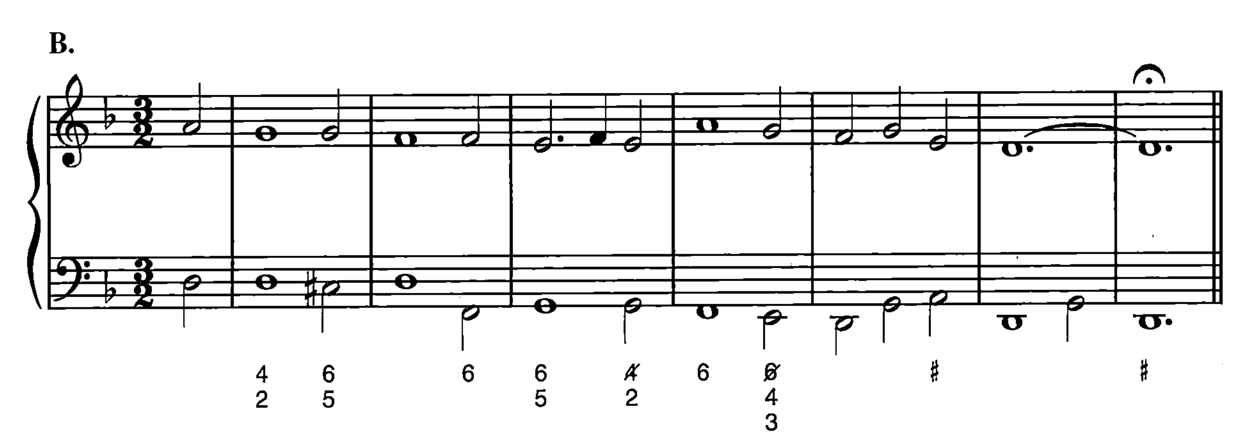

C.

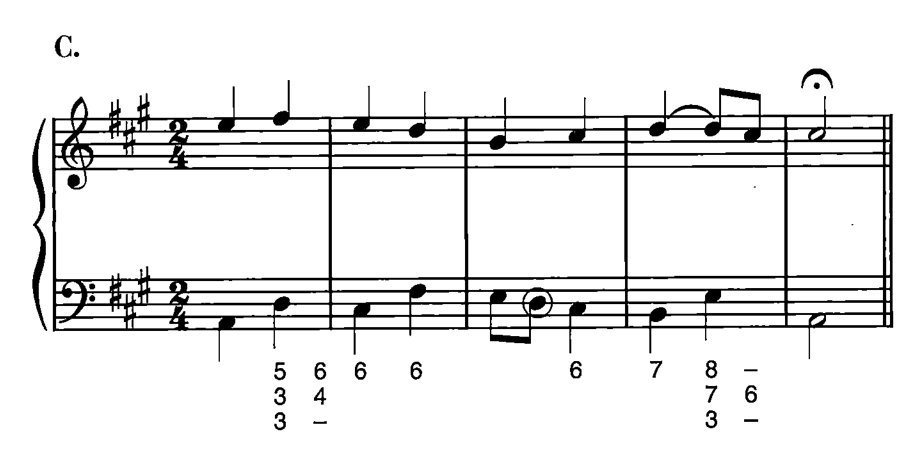

The VI, III, and Other Diatonic Chords

1 Play each of the following progressions in three keys of your choice. Avoid parallels within the deceptive cadence in Exercise 3C.

A. $\hat{3}$ $\hat{3}$ $\hat{2}$ $\hat{2}$ $\hat{1}$
(m) i VI ii° V i

B. $\hat{8}$ $\hat{7}$ $\hat{6}$ $\hat{5}$ $\hat{3}$
(M) I iii IV V I

C. $\hat{8}$ $\hat{8}$ $\hat{8}$ $\hat{7}$ $\hat{8}$
(M) I IV I^6 V vi

D. $\hat{1}$ $\hat{2}$ $\hat{3}$ $\hat{4}$ $\hat{5}$
(m) i VII III iv^6 V

2 Play each progression and then transpose it a major 2nd higher and lower, using both major and minor modes for Exercise 2A.

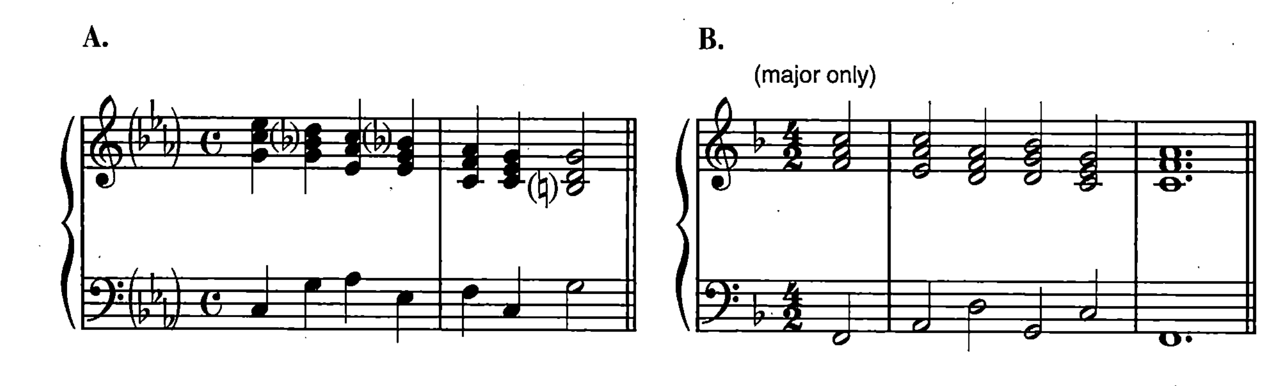

3 Realize the following figured-bass exercises, and be able to analyze them using roman numerals. You must omit the chordal 5th in the last three seventh chords in Exercise 3C to avoid parallels.

A.

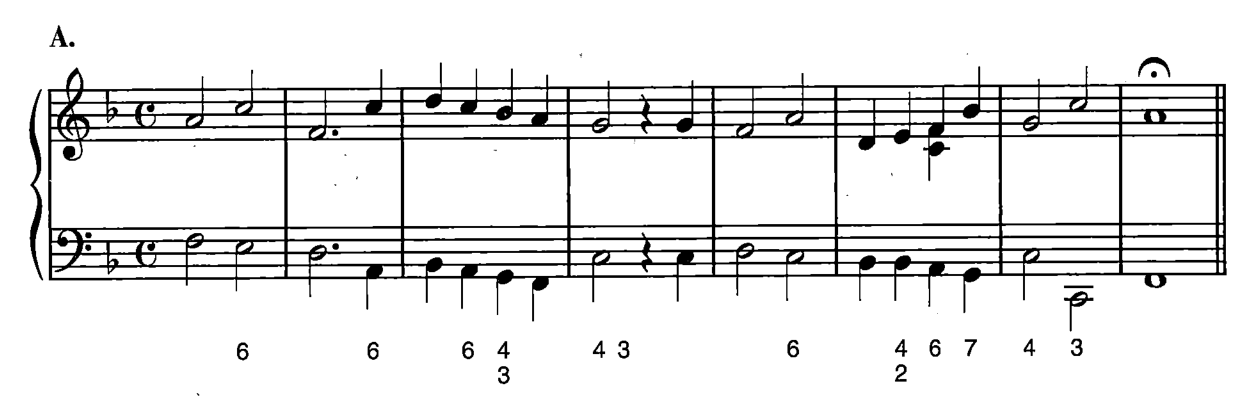

B.

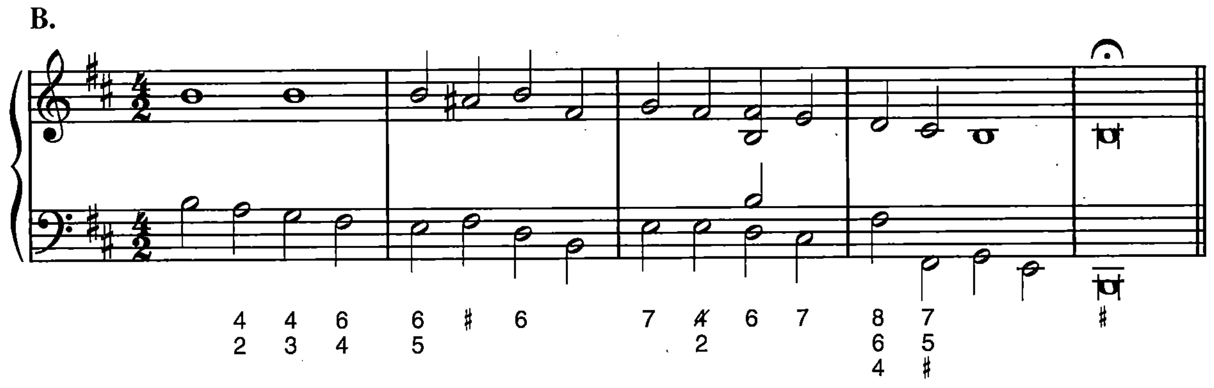

C.

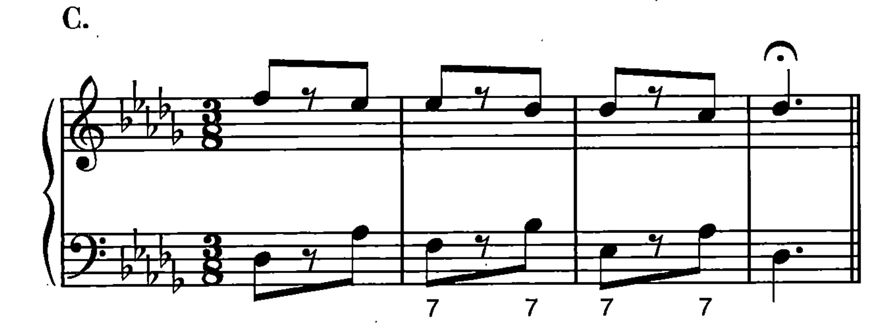

4 Harmonize the following two melodies, utilizing appropriate mediant and submediant chords. For at least one of them, use a texture other than four-voice chorale style.

CHAPTER 19

Tonicization and Modulation I

MOTION TO V AND III

1 The following phrases make a common-chord modulation, either to the dominant [V] (from a major key) or relative major [III] (from a minor key). Harmonize in four-voice texture only the chords that are bracketed, making a cadence in the new key. Use the first chord in each bracket as the pivot.

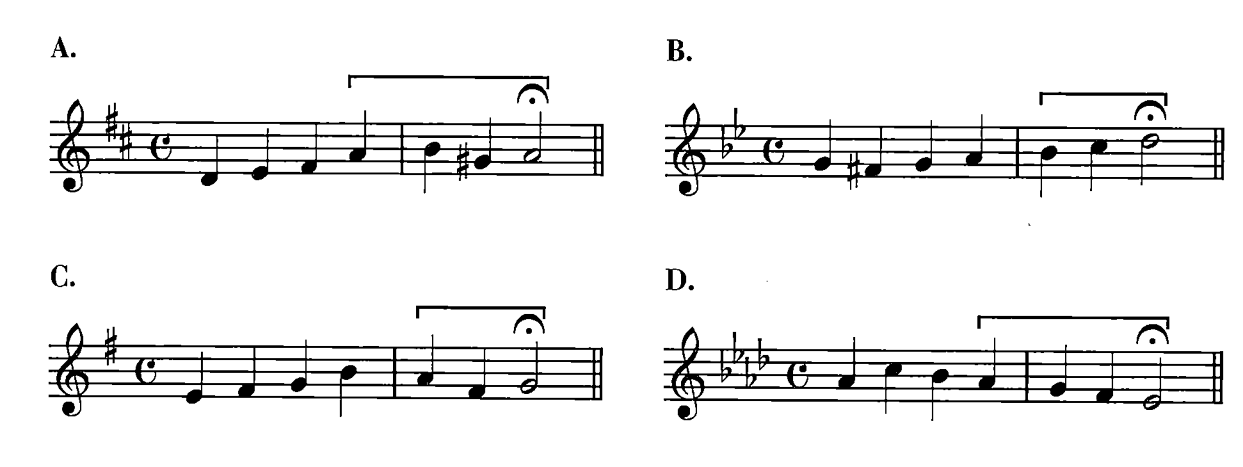

2 Play three separate modulations from a major tonic to its dominant key ([V]) in four-voice texture, using a different pivot chord in each passage. Now do the same thing from a minor tonic to its relative major key ([III]). Use a different key for each modulation. Employ a minimum of five to seven chords. Try to create a smooth, basically stepwise soprano line that leads to an authentic cadence in the new key.

3 Realize the following figured-bass exercises, noting the tonal destination. Recite the roman numerals as you play them, stating where the pivot chord occurs. Do any employ chromatic modulation?

A.

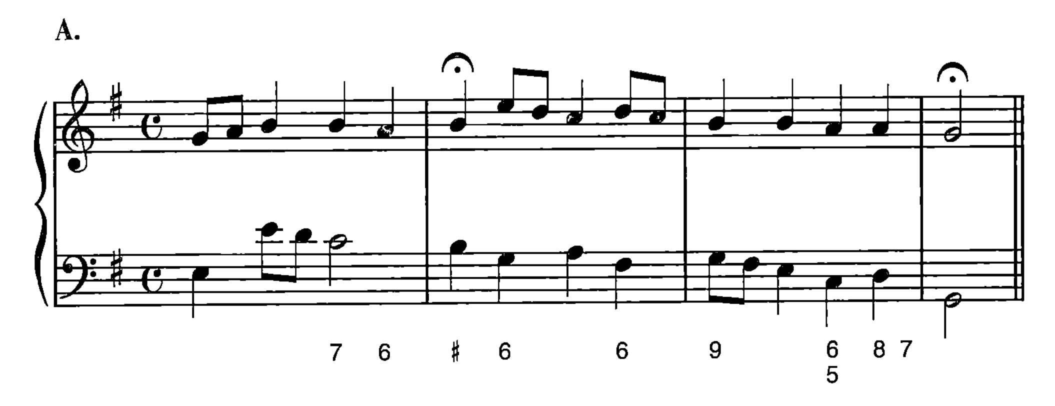

B.

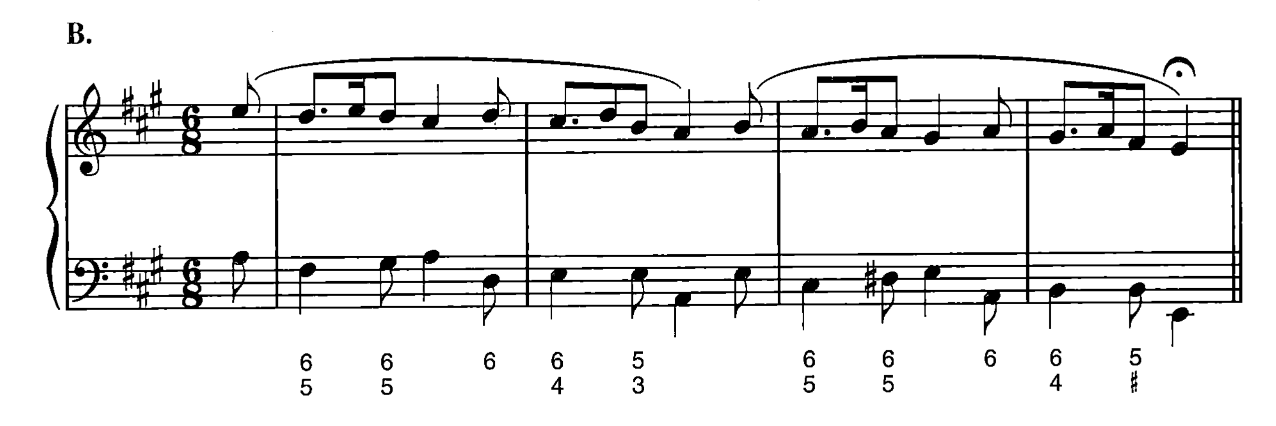

C.

CHAPTER 20

Harmonic Sequences I

ROOT MOVEMENT BY 2ND AND 3RD

1 Continue the following harmonic sequences in strict diatonic fashion. Then transpose each to two other keys of your choice. State why the 7–6 suspensions need to be added in the three-voice Exercise 1C.

A.

B.

C.

2 Realize the following figured-bass exercises, and identify the type of sequence used in each passage.

A.

B.

The Leading-Tone Seventh Chord

1 Play the following chord progressions in minor keys up to four accidentals.

A. $\hat{3}$ $\hat{4}$ $\hat{3}$ — i $\text{vii}^{\circ 7}$ i

B. $\hat{8}$ $\hat{7}$ $\hat{8}$ — i $\text{vii}^{\circ 6}_{5}$ i^{6}

C. $\hat{3}$ $\hat{4}$ $\hat{2}$ $\hat{5}$ — i iv $\text{vii}^{\circ 4}_{3}$ i^{6}

D. $\hat{1}$ $\hat{2}$ $\hat{3}$ $\hat{2}$ — i $\text{vii}^{\circ 4}_{2}$ $\text{V}^{6\ 5}_{4\ \sharp}$

2 Continue the harmonic sequence below, concluding with the key of F♯ minor. You may wish to transpose to another octave. Identify the various inversions of the leading-tone seventh chords.

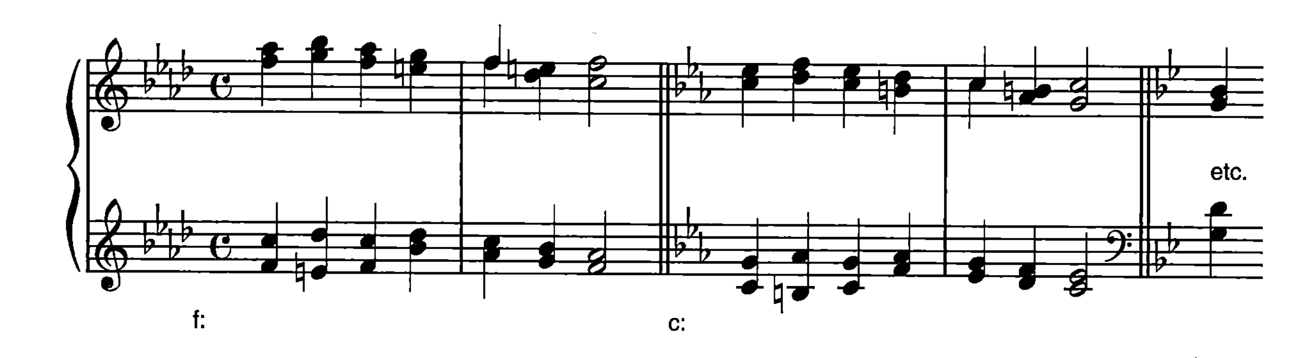

3 Realize the following figured-bass exercises, one or both of which may modulate to another key.

A.

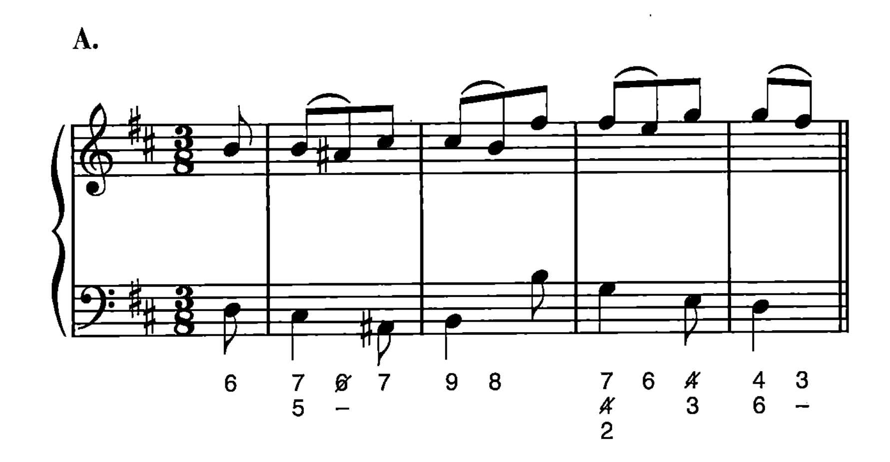

B.

C H A P T E R 2 2

Harmonic Sequences II

ROOT MOVEMENT BY 5TH

1 Play the two cycle-of-fifths progressions below, using the keys of D, F, and A♭ major for the first, and the keys of E, G, and B minor for the second. Why must you alternate complete chords with incomplete chords in Exercise 1B?

A.	$\hat{3}$	$\hat{4}$	$\hat{2}$	$\hat{3}$	$\hat{1}$	$\hat{2}$	$\hat{7}$	$\hat{8}$
(M)	I	IV^6	vii°	iii^6	vi	ii^6	V	I

B.	$\hat{3}$	$\hat{3}$	$\hat{2}$	$\hat{2}$	$\hat{1}$	$\hat{1}$	$\hat{7}$	$\hat{8}$
(m)	i^7	vi^7	VII^7	III^7	VI^7	ii^{ø7}	V^7	i

2 Continue the following sequential patterns to the end of each passage. Then transpose each one to two other keys of your choice.

A.

B.

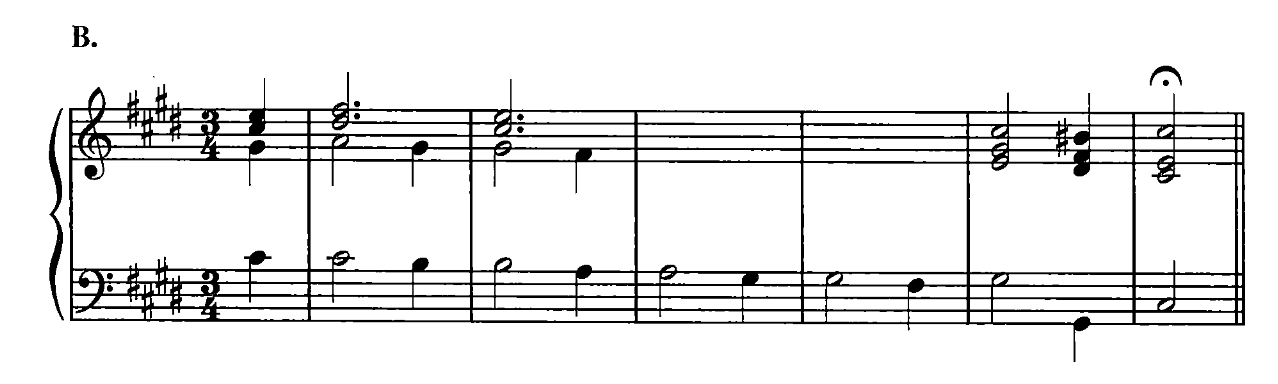

3 Realize the following figured-bass exercises. Determine the basic root movement in each case; some may not move by 5ths.

A.

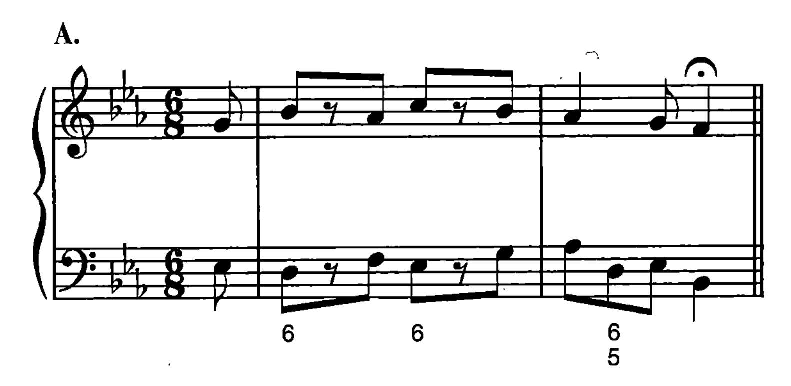

B.

C.

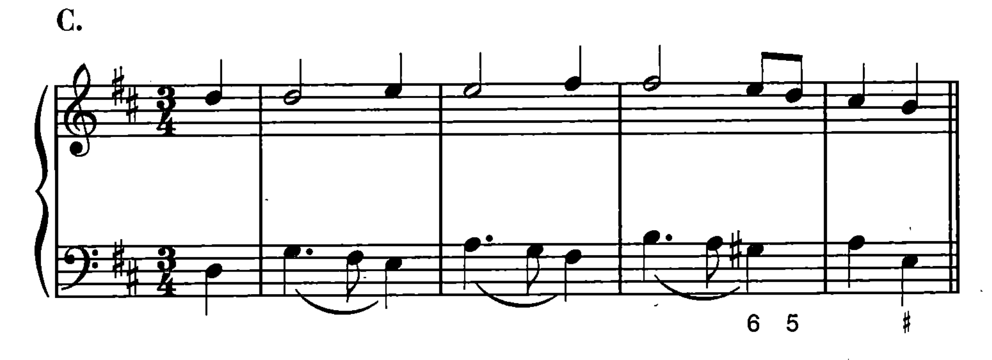

4 Harmonize the tunes below in four-voice texture, using sequential progressions that are implied by the melodies.

C H A P T E R 2 4

Review of Diatonic Harmony

(textbook chapter title: Introduction to Chromatic Harmony)

1 Play the following progressions in the keys of B♭ major, E major, F♯ minor, and C minor. All roman numerals for diatonic chords appear in capital letters; you will have to change some of them to lowercase, depending on the major or minor mode of the key in which you are playing. The initial soprano note is given.

A. $\hat{1}$
I II^4_2 V^6 I I^6 IV V^7 VI

B. $\hat{3}$
I IV^6 V^6 I VI II^6 V I

C. $\hat{1}$
I III IV I IV $V^{6\ 5}_{4\ 3}$ I

Pianists and organists should play one of the above in a freer keyboard texture; non-keyboard majors may continue to use block chords.

2 Continue the following diatonic sequences, and identify the type of sequential motion in each.

A.

B.

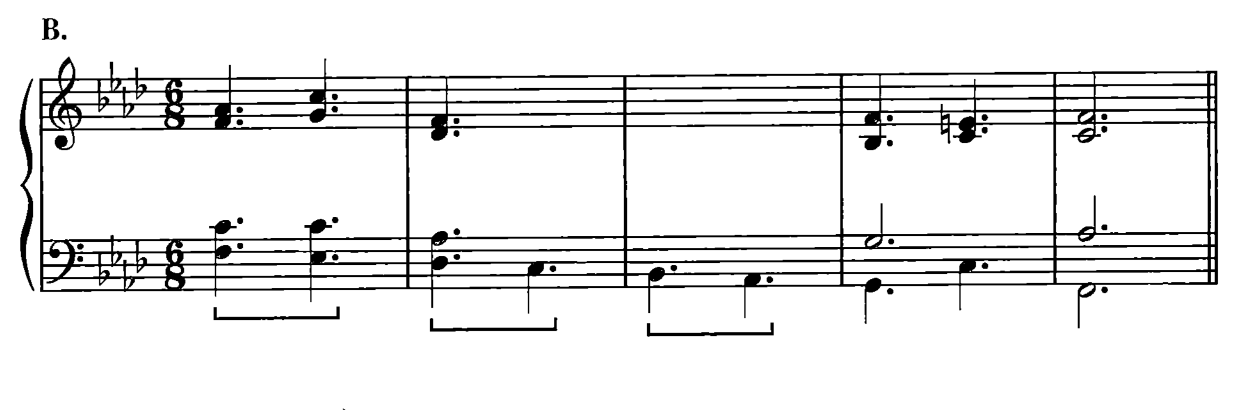

C.

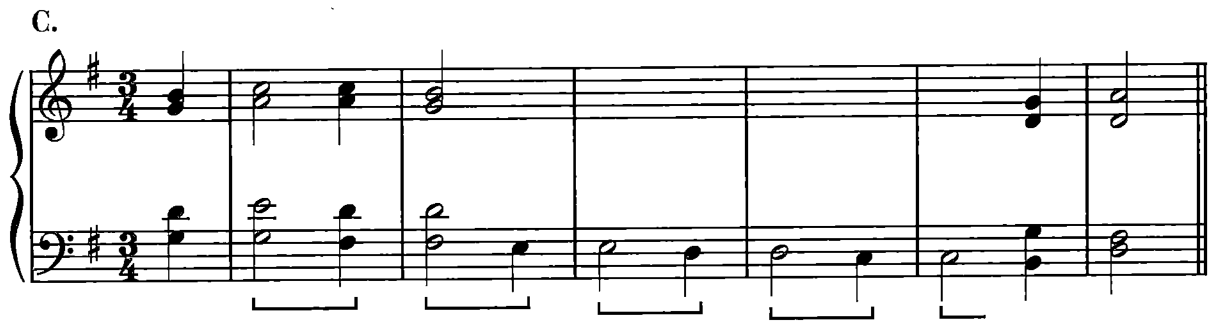

3 Realize the figured-bass chorale below. Use close structure whenever possible. Recite the harmonic functions of the chords as you play them, stating where the modulation occurs.

"Herzliebster Jesu"

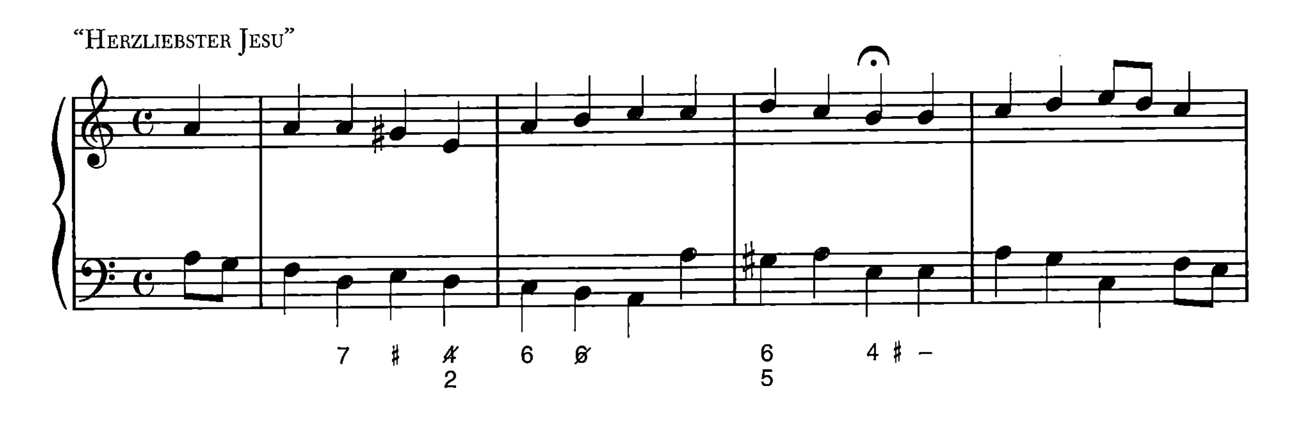

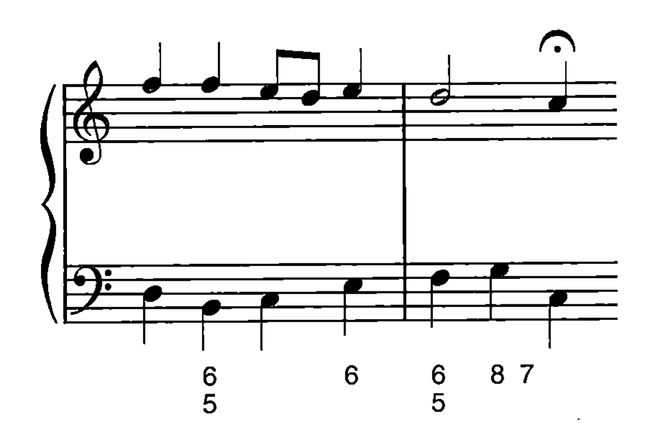

CHAPTER 25

Tonicization and Modulation II

SECONDARY DOMINANTS

1 Play the following harmonic progression and then transpose it to three other keys of your choice. Recite the function of each chord as you play it.

2 Continue the sequence of applied dominant seventh chords in each of the following two progressions, and then transpose each to one other key of your choice. Recite the various chord functions.

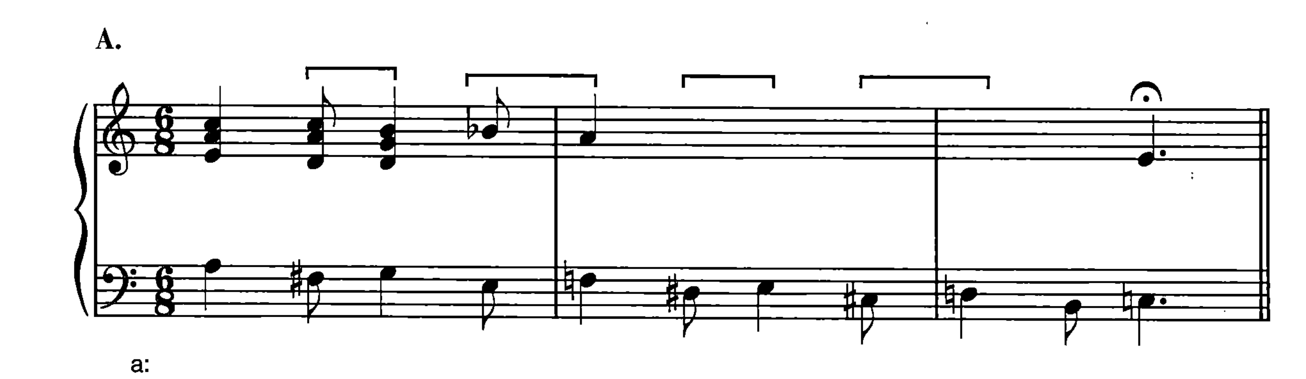

3 Realize the following two figured-bass exercises in four-voice texture, naming each applied chord as you play it.

4 Harmonize the following brief melodic line in either E♭ major or C minor, using applied chords to support those notes that are denoted with an arrow.

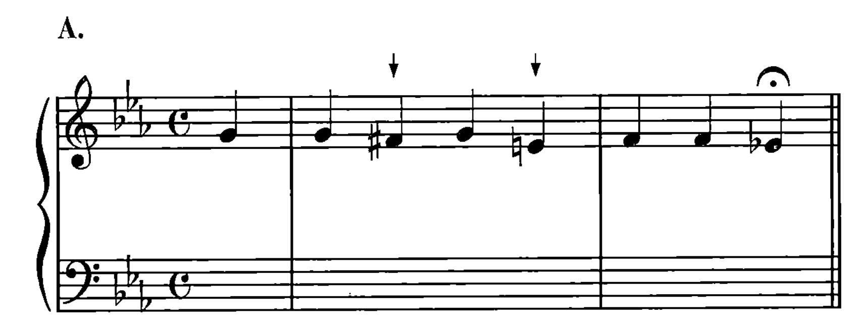

5 Play the chordal succession below, and then transpose it to two other keys of your choice. All of the applied harmonies are diminished seventh chords.

6 Play the following progressions in the keys of C, D, and E. The chords denoted by an *X* represent secondary dominants, some form of either V⁷ or vii°⁷ of the chord that follows. Strive for a smooth bass line, using appropriate inversions, and a good soprano line.

A. Major and minor keys.
I X vi X IV X ii V I
(watch for possible parallels)

B. Minor keys only.
i X VII X VI X V i

7 Realize the following figured-bass exercises, which employ both V⁷ and vii°⁷ applied chords. Analyze the chord functions in each passage.

A.

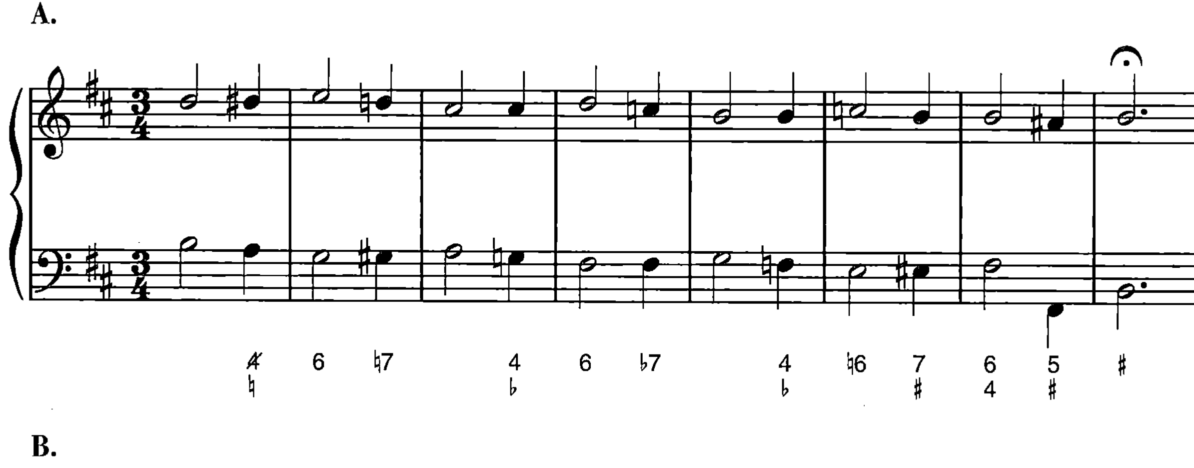

B.

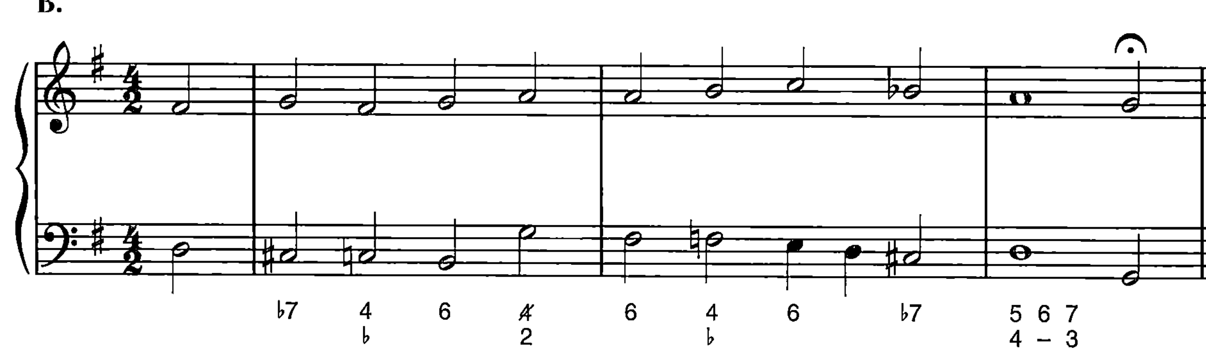

CHAPTER 26

Tonicization and Modulation III

MODULATION TO CLOSELY-RELATED KEYS

1 The following melodic phrases modulate to closely related keys. Play the opening of each phrase, and in four-voice texture, harmonize only the bracketed chords, which form a cadence in the new key. The first bracketed chord will serve as the harmonic pivot; identify its function in both keys.

2 Realize the following figured-bass exercise. What scale degrees are tonicized at each cadence? Are they all closely related to each other *and* to the original tonic key? Recite the chord functions in each passage, pointing out the pivot chords.

3 The following phrases involve a chromatic modulation to a closely related key, as indicated by the key designations within parentheses. The chord that must show the chromatic alteration is denoted with an arrow; place the half-step motion in the bass voice of this chord. Harmonize each phrase in four-voice texture. The progression after the chromatic inflection should establish a cadence in the new key center. Recite the function of your chord choices as you play them. Play Exercises 4B and 4D in both major and minor modes.

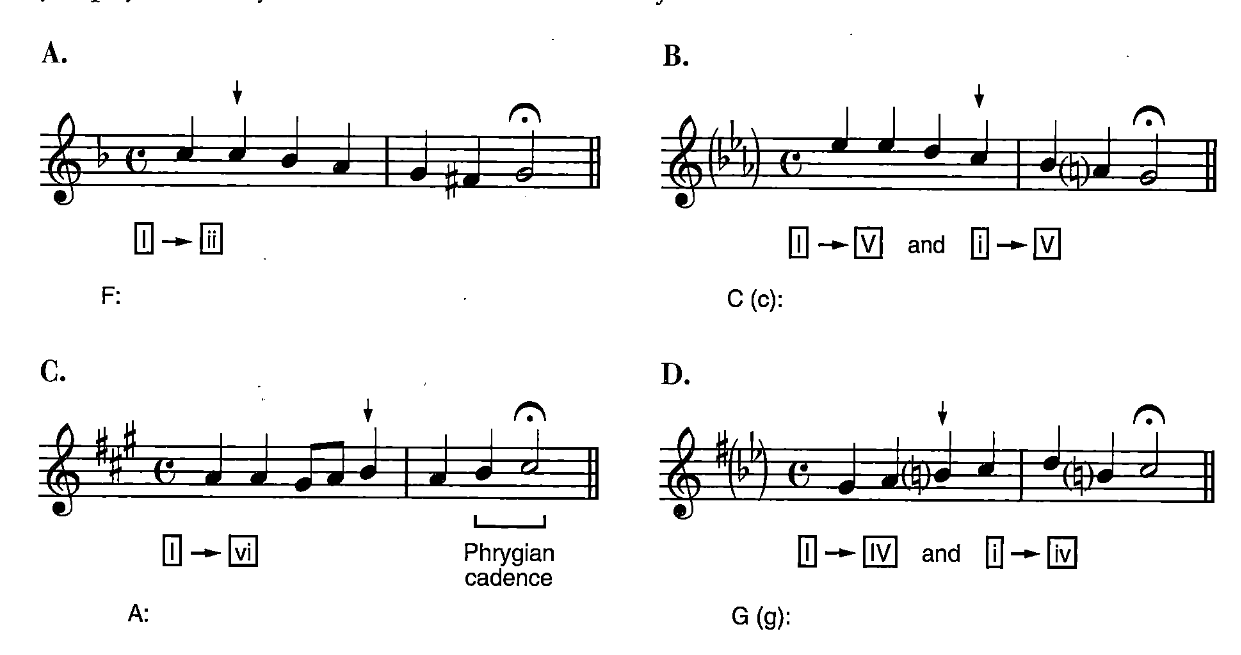

4 Realize the following figured-bass exercise. Recite the chord functions in each passage, stating the means of modulation to the tonicized key areas.

C H A P T E R 2 7

Modal Exchange and Mixture Chords

1 Play a root-position iv, $ii^{ø7}$, ♭VI, ♭III, and III in D major, F major, A major, G major, and B♭ major, using four-voice texture.

2 Play the following voice-leading models, and transpose them to the indicated keys. Recite each mixture chord as you perform it.

A. Transpose to C major.

B. Transpose to B♭ major.

C. Transpose to A major. This passage contains chord progressions that are typical of rock. What type of root movement is employed in the last two measures?

3 Play the following progressions in the keys of D major, E♭ major, A major, and B♭ major. The soprano scale degrees are provided for you.

A. $\hat{5}$ $\hat{4}$ $\hat{4}$ $\hat{3}$ $\hat{3}$ $\hat{4}$ $\hat{2}$ $\hat{3}$
I $\text{ii}^{\varnothing 4}_{2}$ V^{6}_{5} I vi IV $\text{ii}^{\varnothing 6}_{5}$ I

B. $\hat{3}$ $\hat{2}$ $\hat{2}$ $\hat{1}$ $\hat{1}$ $\hat{1}$
I $\text{ii}^{\varnothing 6}_{5}$ V ♭VI iv I

4 Realize the following figured-bass exercises, and recite the chord functions in each passage.

A.

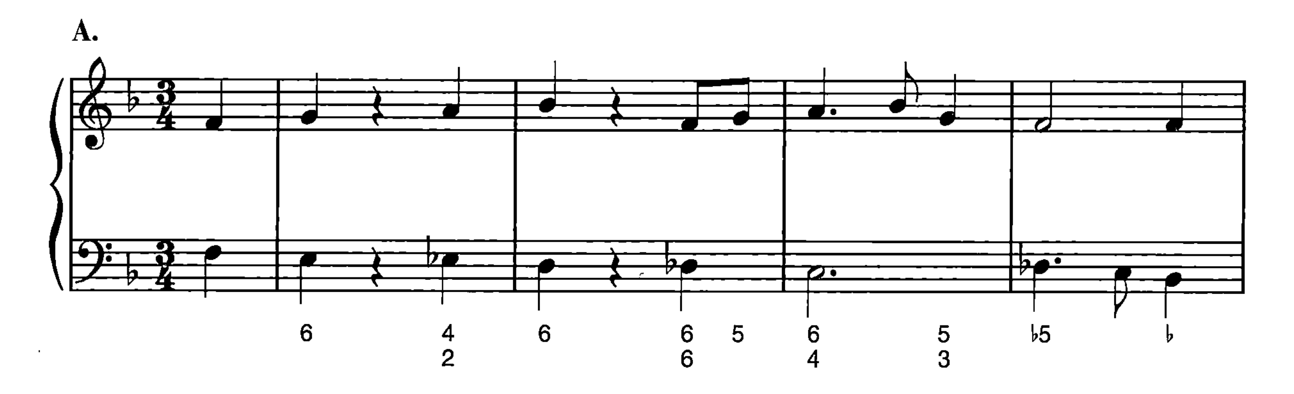

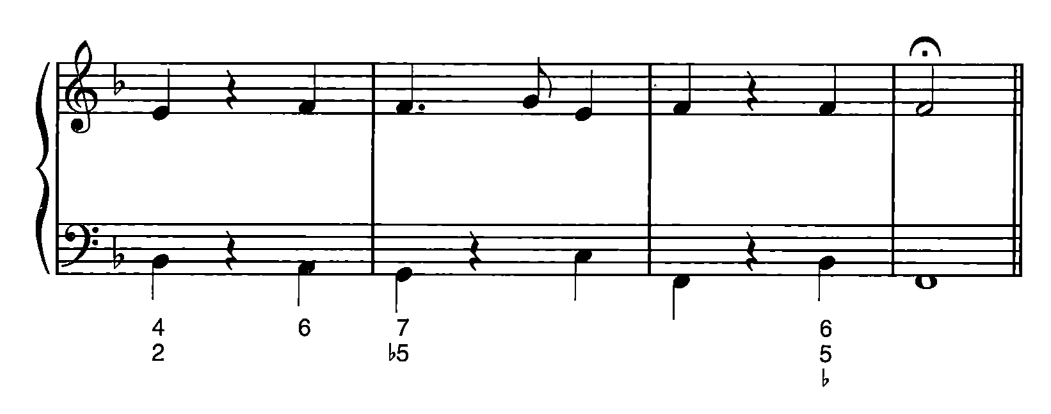

B.

5 Harmonize the following melody, using appropriate mixture harmonies on the notes marked with an arrow.

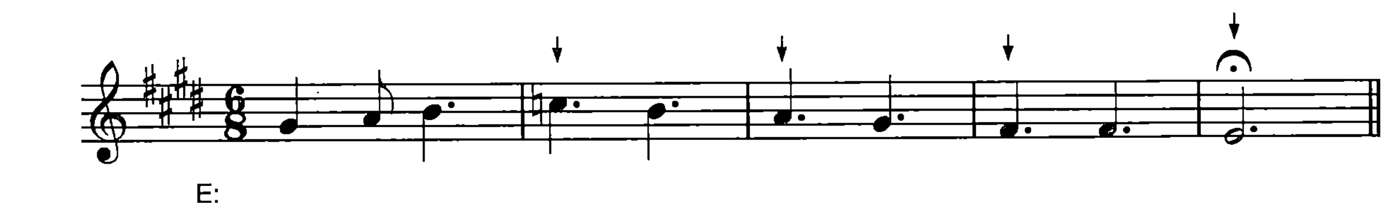

The Neapolitan Chord

1 Play a $\flat II^6$ chord (in four voices) in D minor, B minor, C minor, A minor, and G minor. Try to keep scale degrees $\flat\hat{2}$ or $\hat{4}$ in the soprano, but avoid doubling the altered note.

2 Harmonize the following cadential formulas, using a $\flat II^6$ on those chords indicated with an arrow. Use two different settings for Exercise 2A.

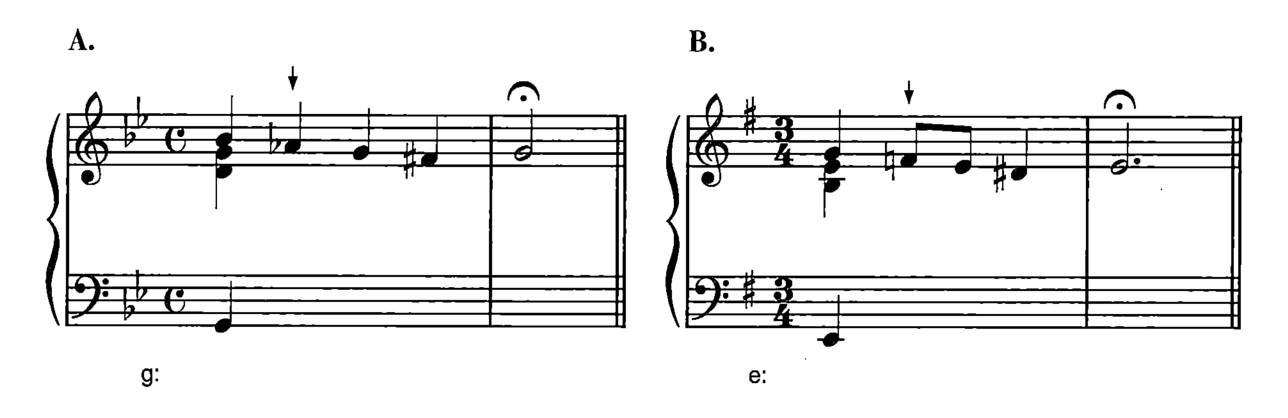

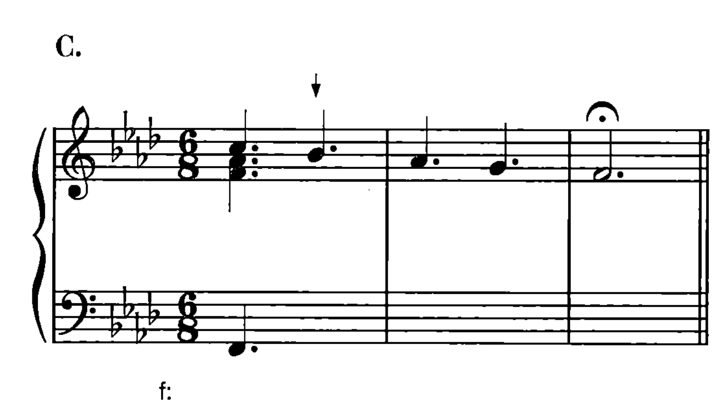

3 The following exercises, most of whose bass notes lack figures, contain various treatments of the Neapolitan harmony. Scan the outer voices to determine the implied chords, and then realize the examples in four-voice texture. Recite the chord functions as you play them; some exercises may contain modulations. Watch for potential parallels in Exercise 3B.

A.

B.

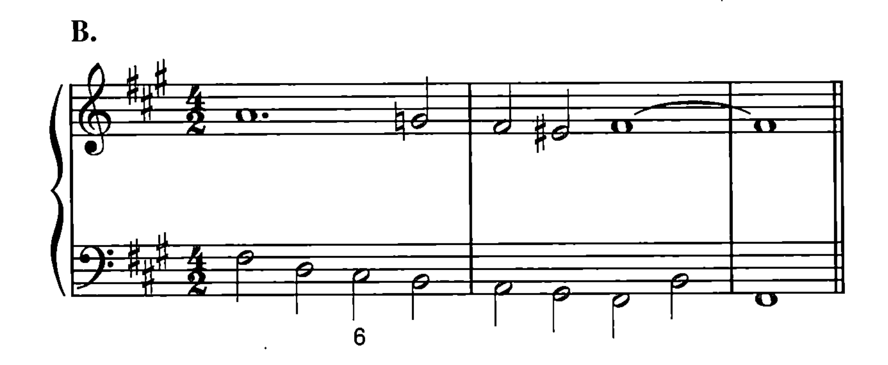

CHAPTER 29

Augmented Sixth Chords

1 Play It^{6}, Ger$^{6}_{5}$, and Fr$^{4}_{3}$ chords in four voices, resolving each to the dominant, in keys up to three accidentals. You may wish to review the resolution of the German Aug6th chord.

2 Play the following cadential progressions, which feature various Aug6th chords. Then transpose them to two other keys of your choice.

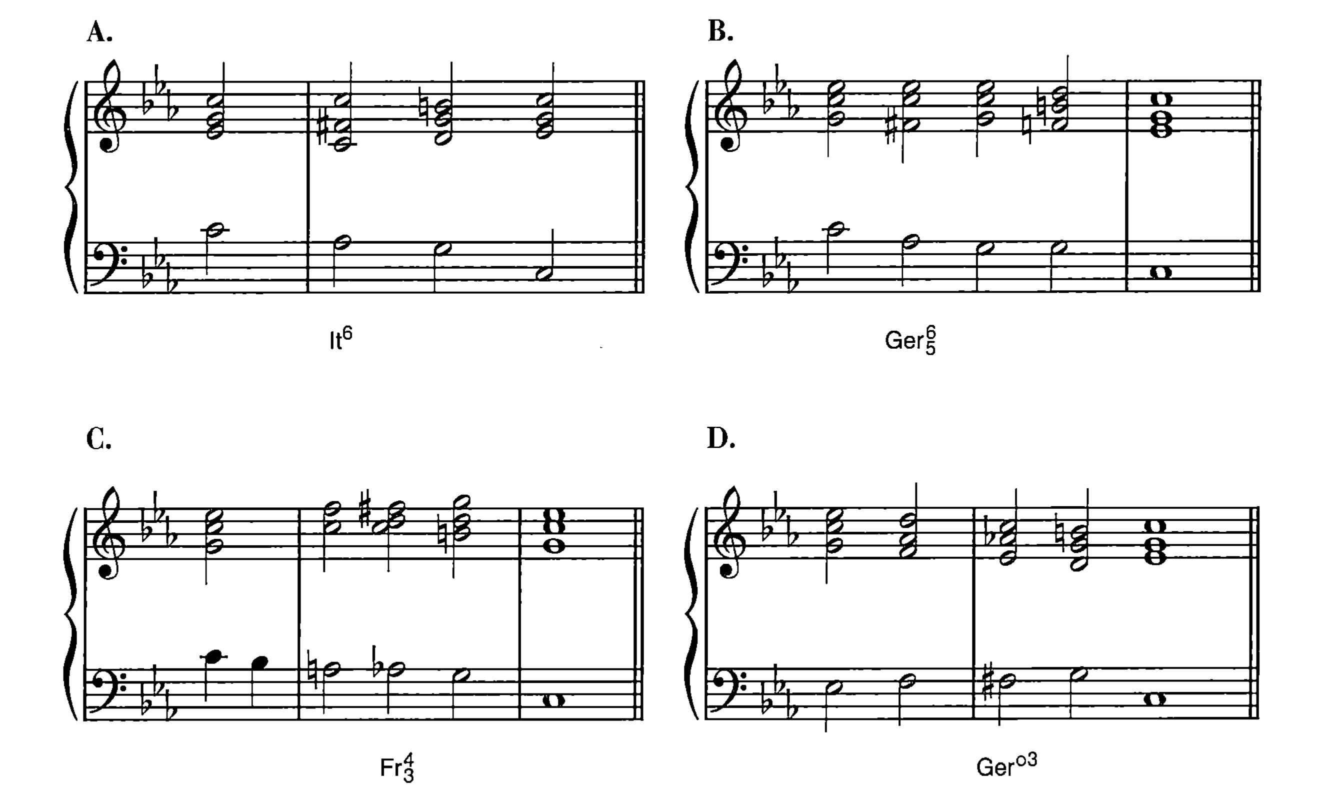

3 Complete the following sequence of secondary It^6 chords.

4 Realize the following figured-bass exercise, naming the Aug6th chords as you play them. Watch for possible key changes.

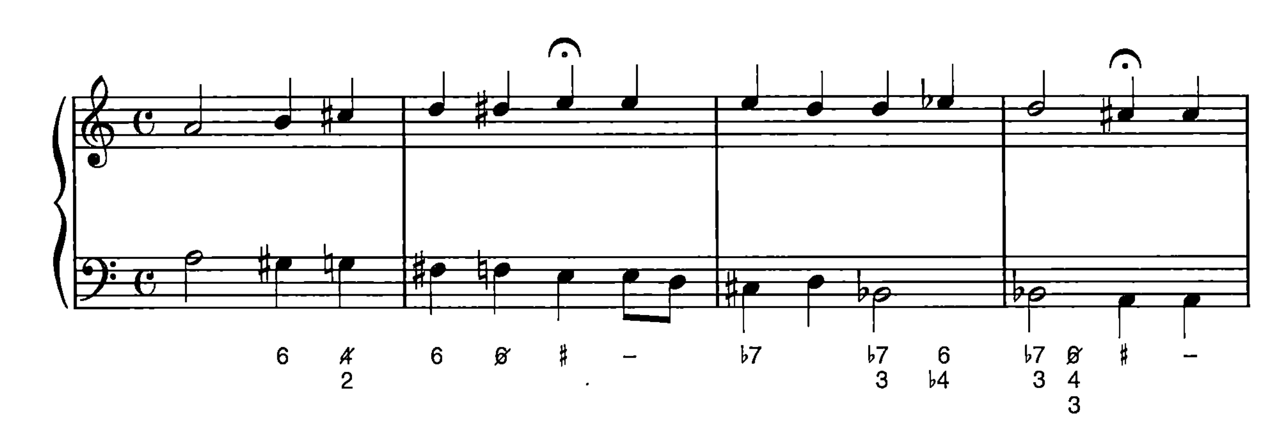

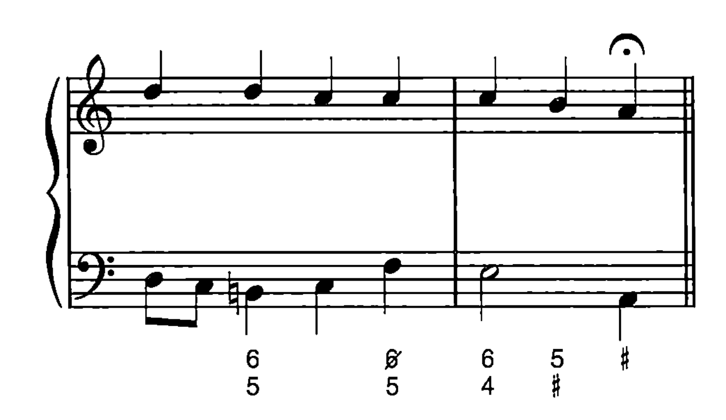

CHAPTER 31

Ninth, Eleventh, Thirteenth, and Added-Note Chords

1 The following root-position chords on E are denoted by commercial symbols. Transpose each chord to the roots G, A, D, F, and C. (You may wish to refer to Appendix 4.)

E^9 $E^{9(MAJ7)}$ E^{-9} Em^9 E^{11} (omit the 3rd) Em^{11} E^{+11}

2 Approach and resolve either a V^9 (in a major key) or V^{-9} (in a minor key) in four-voice texture. What chord member must you omit in the ninth chords?

3 Play the following progressions, and then transpose them to two other keys of your choice. Identify the chord type of each sonority as you play it.

A.

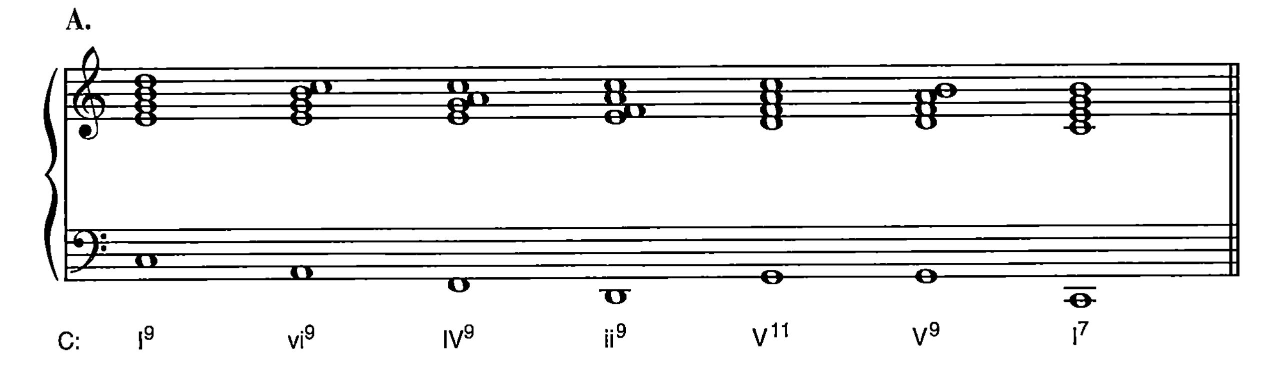

B.

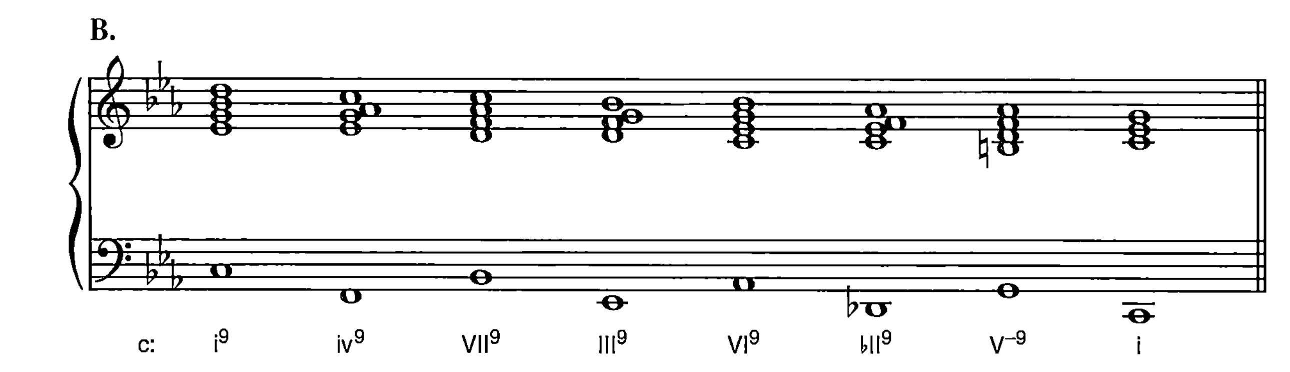

4 Play the indicated chords above the given bass, using *five*-voice texture. Make sure your soprano line moves primarily by step. In this assignment, you need not be concerned with parallels since this passage is in jazz style.

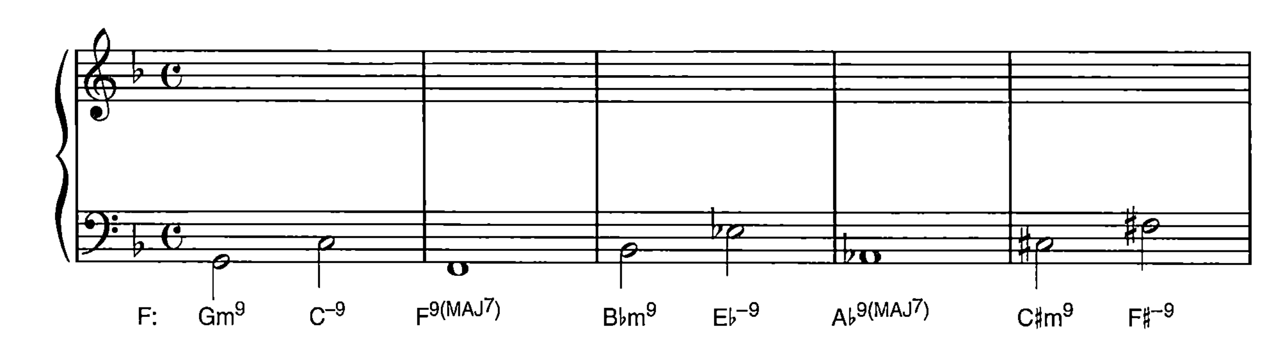

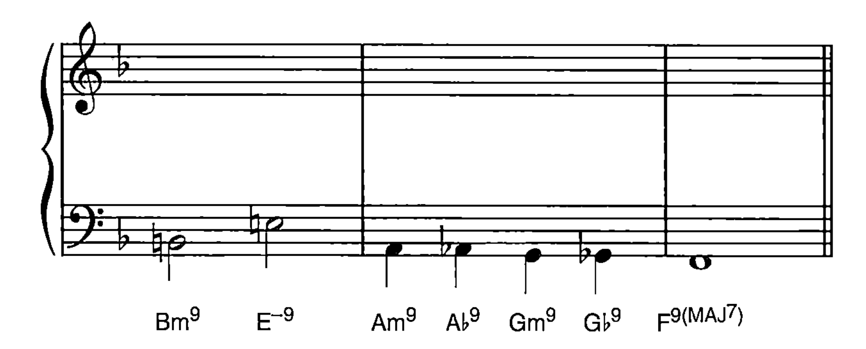

C H A P T E R 3 2

Embellishing Chromatic Chords

1 Play the following passage and then transpose it to two keys of your choice. Cite all examples of augmented triads as you play them, and identify their linear derivation.

2 The following two cadences employ altered forms of the V^7. Play each in major keys up to and including three accidentals.

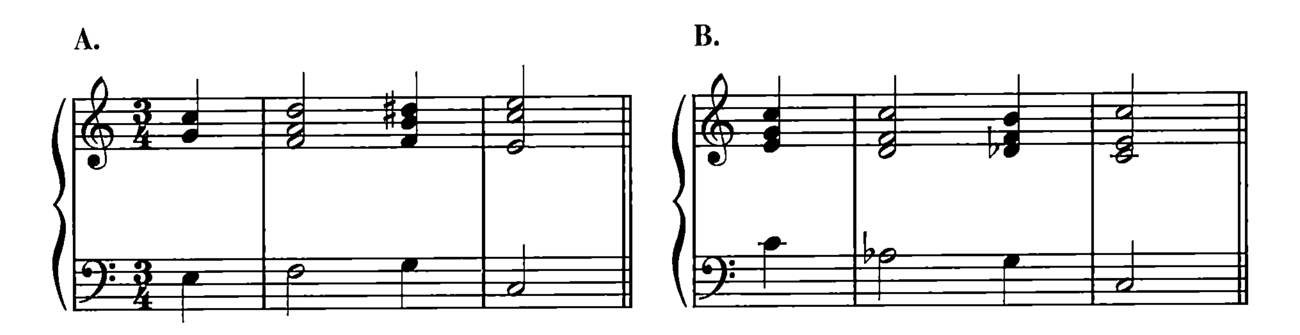

3 Carefully realize the following figured-bass exercises, which are somewhat more demanding than in previous chapters. Recite the various embellishing chords and explain their melodic derivation.

C H A P T E R 3 4

Modulation to Foreign Keys I

1 Realize the following two figured-bass passages. Identify the foreign key relation and pivot chord in any modulation away from and back to the original key center. The figuration is rather complicated here; remember that it always refers to the intervals above the bass note in terms of the given key signature.

2 Choosing any major key as your point of departure, play a short progression to establish the key, then modulate to three different foreign key areas—♭II, vii, and VI—returning to the original key after each modulation. In the last modulation, use an altered pivot chord or perhaps a common tone. Be sure to conclude with a confirming cadence in each new key.

CHAPTER 35

Harmonic Sequences III

CHROMATIC ELABORATIONS OF DIATONIC SEQUENCES

1 Play the three different settings of the chromatic tetrachord given below, and transpose each of them to two other keys of your choice. Cite the type of underlying sequence in each case.

A.

B.

2 Play the beginning of each sequential passage, and continue the sequence to the concluding cadence. The sequence in Exercise 2A will be somewhat free.

CHAPTER 37

Chromatic Voice Leading

1 Play the following C-major pattern (mm. 1–2), which employs chromaticism by contrary motion, and then transpose it by successive 5ths, starting with F major. The beginning of each pattern is given. What harmony is being prolonged in each case?

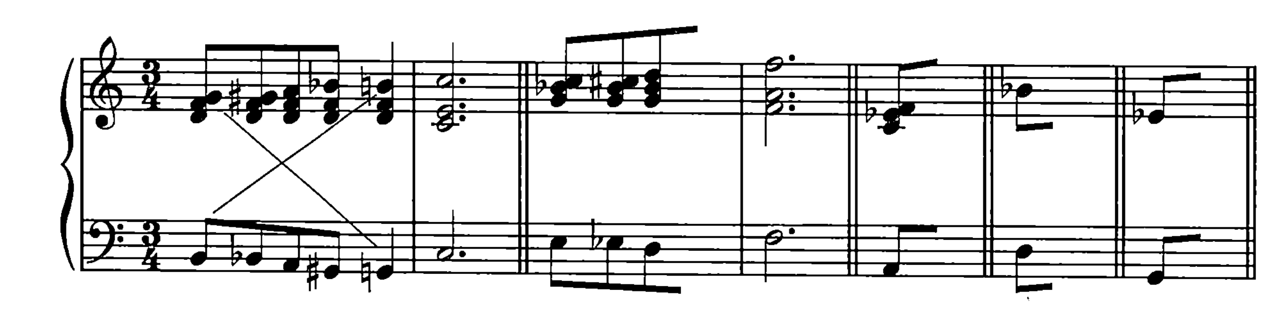

2 Play the following passage, and then transpose it to one other key of your choice. Identify the chromatic voice leading as you play.

3 Realize the following figured-bass progressions. In Exercise 3B, the figuration is rather complicated. Cite any instances of chromaticism by contrary motion that you notice.

Sometimes composers spell out their names in musical notation. Whose name is spelled by the first four pitches in the bass of Exercise 3B? (You may wish to consult chapter 40 of the text.)

A.

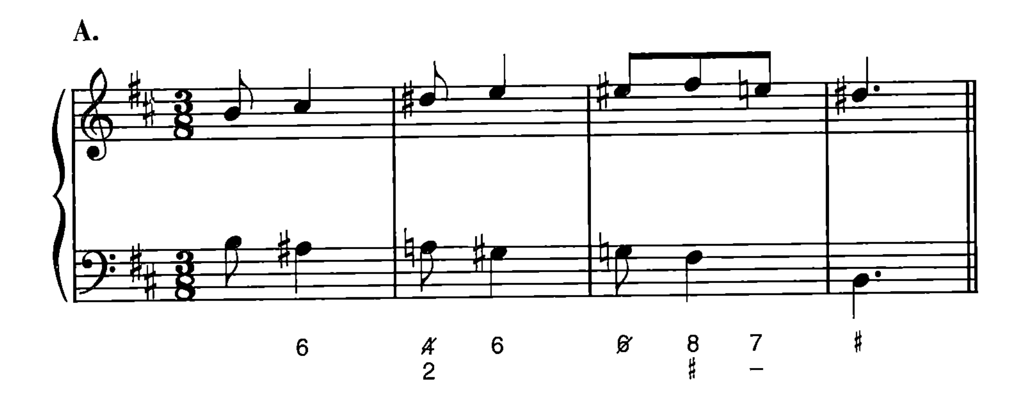

B.

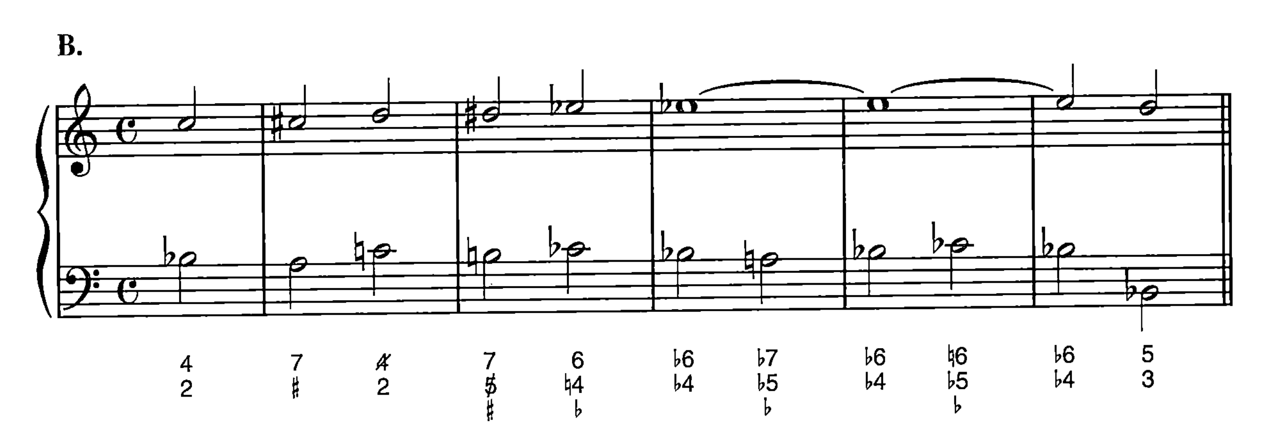

C H A P T E R 3 8

Modulation to Foreign Keys II

1 Use the diminished-seventh chord F♯ A C E♭ in the following passage as a pivot chord (spelled enharmonically, if necessary) to modulate from C minor to G minor, B♭ minor, C♯ minor, and E minor, respectively. Be sure to conclude with a confirming cadence in your new key.

2 Using a Ger$^6_5$ spelled enharmonically as a V^7, or vice versa, compose three short chord progressions that modulate from C minor to D♭ major, C minor to B minor, and C minor to E minor. Be sure to establish both your original key and the new key each time.

3 Continue the following half-step sequence to B♭ major. How is the series of ascending chromatic tonicizations achieved?

4 The following pair of passages shift to a foreign key by strict sequential modulation. Play the original phrase, and then follow it with a sequential repetition in the indicated key.

A.

B.

C H A P T E R 3 9

Symmetrical Divisions of the Octave

1 Continue the following two sequences in strict fashion. In Exercise 1B, the pattern will move from C major through A♭ and E and then back to the original key. What is the function of the even-numbered chords in the first passage?

A.

B.

2 This passage prolongs a V^7 through an omnibus progression by descending minor 3rds. Play and then transpose it to two other keys of your choice.

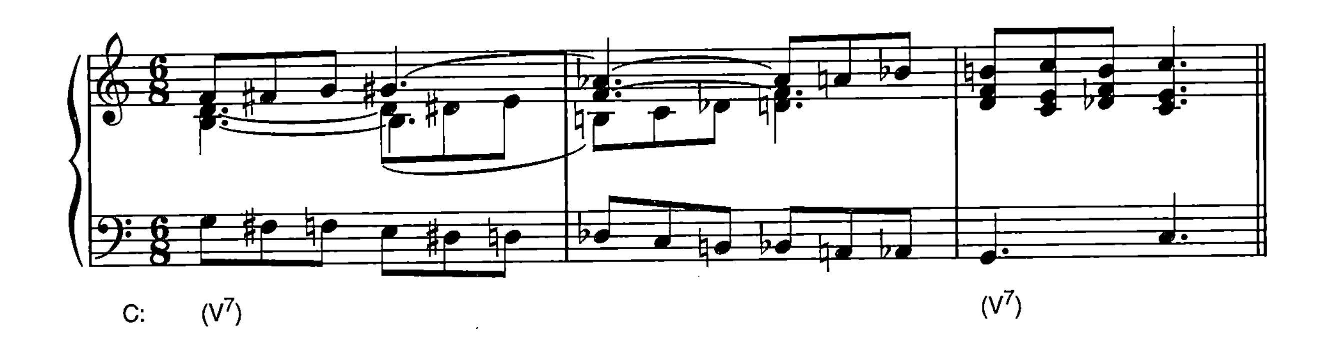

3 The following exercise is based on the opening section of the "Liebestod" from Wagner's *Tristan und Isolde.* Realize the figured bass in the first two measures, and then continue the sequence in strict fashion. What scale is suggested by the notes indicated with arrows in the soprano line? (You may wish to refer to the end of Appendix 2.)